Latest Research in Contaminants of Emerging Concern

Edited by Davide Di Paola and Caterina Faggio

Published in London, United Kingdom

Latest Research in Contaminants of Emerging Concern
http://dx.doi.org/10.5772/intechopen.1008665
Edited by Davide Di Paola and Caterina Faggio

Assistants to Editors: Cristiana Roberta Multisanti and Federica Impelliteri

Contributors
Aisyah Farhanah Abdul Majid, Alexis Rubén Bracamontes-Ruelas, Anwar Usman, Caterina Faggio, Cristiana Roberta Multisanti, Davide Di Paola, Federica Impellitteri, Gerson Lira-López, Harry Lik Hock Lau, Lucia del Pilar Rodríguez-Hernández, Mahdi Banaee, Michael Greenberger, Muhammad Nur, Nur Aisyah Abdul Munir, Nur Diana Bazilah Awang Idris, Nur Nabaahah Roslan, Ricardo Herrera-Navarrete, Rusydi R. Sofian, Shirley Maloney Mueller, Syahirah Nabilah Aedy Aewandy, Victor Samir Jaik-Morán, Ángeles Martínez-Orgániz

First published in London, United Kingdom, 2026 by IntechOpen
IntechOpen is the global imprint of INTECHOPEN LIMITED, registered in England and Wales, registration number: 11086078, 167-169 Great Portland Street, London, W1W 5PF, United Kingdom

For EU product safety concerns: IN TECH d.o.o., Prolaz Marije Krucifikse Kozulić 3, 51000 Rijeka, Croatia, info@intechopen.com or visit our website at intechopen.com.

British Library Cataloguing-in-Publication Data
A catalogue record for this book is available from the British Library

Latest Research in Contaminants of Emerging Concern
Edited by Davide Di Paola and Caterina Faggio
p. cm.

This title is part of the Environmental Sciences Book Series, Volume 49
Topic: Pollution
Series Editor: J. Kevin Summers
Topic Editors: Ismail Rahman and Zinnat Ara Begum

Print ISBN 978-1-80631-058-6
Online ISBN 978-1-80631-057-9
eBook (PDF) ISBN 978-1-80631-059-3
ISSN 2754-6713

If disposing of this product, please recycle the paper responsibly.

IntechOpen Book Series

Environmental Sciences

Volume 49

Aims and Scope of the Series

Scientists have long researched to understand the environment and man's place in it. The search for this knowledge grows in importance as rapid increases in population and economic development intensify humans' stresses on ecosystems. Fortunately, rapid increases in multiple scientific areas are advancing our understanding of environmental sciences. Breakthroughs in computing, molecular biology, ecology, and sustainability science are enhancing our ability to utilize environmental sciences to address real-world problems.

The four topics of this book series - Pollution; Environmental Resilience and Management; Ecosystems and Biodiversity; and Water Science - will address important areas of advancement in the environmental sciences. They will represent an excellent initial grouping of published works on these critical topics.

Meet the Series Editor

J. Kevin Summers has been working as a Senior Research Ecologist at the Environmental Protection Agency's (EPA) Gulf Ecosystem Measurement and Modeling Division. He worked with colleagues in the Sustainable and Healthy Communities Program to develop an index of community resilience to natural hazards, an index of human well-being that can be linked to changes in the ecosystem, social and economic services, and a community sustainability tool for communities with populations under 40,000. He led research efforts for indicator and indices development. Dr. Summers is a systems ecologist and began his career at the EPA in 1989 and has worked in various programs and capacities. This includes leading the National Coastal Assessment in collaboration with the Office of Water, which culminated in the award-winning National Coastal Condition Report series (four volumes between 2001 and 2012), and which integrates water quality, sediment quality, habitat, and biological data to assess the ecosystem condition of the United States estuaries. He was acting National Program Director for Ecology for the EPA between 2004 and 2006. He has authored approximately 150 peer-reviewed journal articles, book chapters, and reports, and has received many awards for technical accomplishments from the EPA and from outside of the agency. Dr. Summers holds a BA in Zoology and Psychology, an MA in Ecology, and Ph.D. in Systems Ecology/Biology.

Meet the Volume Editors

Davide Di Paola is a researcher at the University of Messina, Italy, where his work focuses on environmental toxicology, aquatic biology, and biomarker-based assessment of contaminant exposure. His research investigates the physiological and molecular mechanisms underlying pollutant effects in aquatic organisms, contributing to the understanding of ecosystem health and environmental risk assessment. He has participated in several national and international research projects and has published widely in peer-reviewed journals. His academic interests include environmental pollution, ecotoxicology, and trophic ecology.

Professor Caterina Faggio is currently a Full Professor of Physiology in the Department of Chemical, Biological, Pharmaceutical and Environmental Sciences at the University of Messina, Italy. She teaches in various bachelor's and master's degree courses at the University of Messina. Her research explores the effects of emerging contaminants on aquatic organisms, with a focus on cellular and biochemical biomarkers. She has authored numerous scientific publications and book chapters and has served as a reviewer and editor for several international journals. Professor Faggio's work emphasizes integrating environmental physiology and toxicology to develop sustainable environmental protection strategies.

Contents

Preface

Environmental contamination has emerged as one of the defining scientific and societal challenges of the twenty-first century. The increasing complexity of pollutants and their interactions within ecosystems calls for interdisciplinary approaches that bridge biology, chemistry, toxicology, and environmental engineering. This edited volume brings together recent insights and conceptual advances from leading researchers, aiming to provide a comprehensive overview of emerging environmental contaminants and their biological implications.

The opening chapter, "Biomarkers of Pesticide Exposure in Aquatic Organisms: Mechanisms of Toxicity and Early Warning Indicators", explores the molecular and physiological responses of aquatic species to pesticide exposure. It highlights the growing use of biomarkers as tools for early detection of ecological stress and for understanding the mechanistic pathways of toxicity. By elucidating how pesticides disrupt metabolic and cellular processes, this chapter sets the scientific foundation for assessing ecosystem health in contaminated waters.

In "Microplastics in Marine and Coastal Ecosystems: Sources, Distribution, Ecological Impact, and Mitigation Prospects", the authors examine one of the most pervasive pollutants of modern times. The chapter traces the sources and environmental fate of microplastics, describes their interactions with marine organisms, and evaluates ongoing efforts to mitigate and manage them. The synthesis of ecological, chemical, and policy perspectives makes this contribution particularly valuable for researchers and practitioners seeking integrated solutions.

The third chapter, "Pharmaceuticals as Emerging Contaminants in the Environment: Impacts and Remediation Processes", focuses on the growing concern over pharmaceutical residues in soil and water systems. It provides an overview of their pathways of entry into the environment, potential ecological and human health effects, and the latest developments in remediation technologies. The discussion emphasizes the need for sustainable strategies to manage pharmaceutical pollution at both local and global scales.

Following this, "Pharmaceuticals in Wastewater: Current Issues and Future Trends for Their Treatment — A Critical Review" deepens the exploration of pharmaceutical contaminants by focusing specifically on wastewater treatment. The authors critically analyze the efficiency of existing treatment technologies, identify research gaps, and propose directions for future development. This chapter complements the preceding one by linking environmental science with technological innovation and policy implementation.

The volume concludes with "Tea, Teapots, and Toxicity in the Seventeenth and Eighteenth Century", by Shirley Maloney Mueller and Michael Greenberger. This historical perspective offers a fascinating counterpoint to the scientific chapters, illustrating how concerns about toxicity and material safety have deep roots in human history. An examination of cultural practices and the scientific understanding of earlier centuries reminds readers that the interplay between human innovation and environmental risk is not a recent phenomenon.

Collectively, the chapters in this volume highlight the dynamic nature of environmental contamination research, spanning molecular mechanisms, ecological impacts, technological responses, and historical insights. The editors hope that this collection will serve as a valuable reference for researchers, students, and professionals engaged in environmental sciences and will inspire continued collaboration across disciplines in addressing the pressing environmental challenges of our time.

The editors would like to express their gratitude to all contributing authors for their expertise and dedication. Special thanks are extended to the editorial and production teams for their tireless efforts throughout the preparation of this volume.

Davide Di Paola and Caterina Faggio
University of Messina,
Messina, Italy

Chapter 1

Biomarkers of Pesticide Exposure in Aquatic Organisms: Mechanisms of Toxicity and Early Warning Indicators

Davide Di Paola, Cristiana Roberta Multisanti, Mahdi Banaee, Federica Impellitteri and Caterina Faggio

Abstract

Pesticides are widely considered a major class of xenobiotics that pose threats to aquatic ecosystems. Their widespread use in agriculture and urban settings results in frequent detection in surface waters, leading to sublethal and chronic effects in aquatic organisms. This chapter reviews the types and sources of pesticides in aquatic environments, their modes of action and biotransformation in aquatic organisms, and the array of biomarkers that can detect early biological responses. Particular attention is given to oxidative stress, endocrine disruption, immunotoxicity, and behavioral alterations. These biomarkers offer valuable and effective tools for ecological risk assessment and environmental health monitoring.

Keywords: pesticides, aquatic toxicology, biomarkers, oxidative stress, endocrine disruptors

1. Introduction

Pesticides are one of the most extensively used groups of xenobiotics in modern society, designed primarily to manage and eliminate agricultural, domestic, and vector-borne pests. While their efficacy in improving agricultural productivity and disease control is undeniable, the unintended ecological consequences, particularly within aquatic environments, have become a growing concern in environmental toxicology and public health research. The continuous application and unregulated runoff of these chemicals have led to their persistent detection in freshwater systems, estuaries, and marine ecosystems, where they can adversely affect nontarget organisms across multiple trophic levels [1–5].

Aquatic ecosystems are particularly vulnerable to pesticide contamination due to multiple exposure pathways, including surface runoff, spray drift, leaching, and discharge from wastewater treatment plants. These processes contribute to chronic low-level exposure and occasional acute contamination events that can alter community structure, impair physiological processes, and lead to long-term population declines [6–8]. In particular, the bioaccumulation potential and sublethal effects of many pesticides make them critical stressors that must be addressed through comprehensive monitoring strategies.

In this context, biomarkers have emerged as vital tools in ecotoxicology for assessing early biological responses to pesticide exposure [9]. Biomarkers are measurable changes in biological systems, ranging from molecular and cellular responses to behavioral and physiological alterations, which can provide insight into the mechanisms of toxicity and potential ecological consequences [10]. Their value lies in their sensitivity, specificity, and ability to detect sublethal and preclinical effects before population-level impacts manifest.

The selection and validation of appropriate biomarkers require a deep understanding of pesticide chemistry, toxicokinetics, and toxicodynamics. Modern analytical approaches now allow researchers to measure a wide range of biomarkers, from classical enzymatic responses such as acetylcholinesterase (AChE) inhibition to advanced omics-based indicators that capture gene expression and metabolic shifts [11,12]. These advances enhance our ability to evaluate the ecological risks posed by complex pesticide mixtures and understand their cumulative and synergistic effects [13,14].

This chapter aims to provide a comprehensive overview of the biomarker responses to pesticide exposure in aquatic organisms, emphasizing a mechanistic understanding and practical applications in environmental monitoring. This chapter explores the classification and environmental behavior of pesticides, the mechanisms of biotransformation and detoxification, and the various biomarkers used at different levels of biological organization. It also integrates recent developments in omics technologies and microbiome science, offering a future-oriented perspective on the role of biomarkers in environmental risk assessment.

2. Pesticides as xenobiotics in aquatic ecosystems

2.1 Classification of pesticides: Chemistry, functionality, and side effects

Pesticides encompass a chemically and functionally diverse array of compounds designed to target specific pest groups; yet, many exhibit nonselective toxicity that extends to aquatic organisms. They can be broadly classified based on their chemical structure and target organism:

- **Organophosphates** (e.g., chlorpyrifos, diazinon, malathion) are highly toxic neurotoxins that irreversibly inhibit AChE. Despite their relatively short environmental half-lives, their acute effects on nontarget species make them a major ecotoxicological concern [15–18].

- **Carbamates** (e.g., carbaryl, aldicarb) share a similar mechanism with organophosphates but exhibit reversible AChE inhibition. Their toxicokinetics in aquatic species are well-documented [19].

- **Pyrethroids** (e.g., permethrin, deltamethrin, cypermethrin) are synthetic analogs of natural pyrethrins. These lipophilic compounds disrupt voltage-gated sodium channels and are highly toxic to aquatic invertebrates and fish [20–22].

- **Neonicotinoids** (e.g., imidacloprid, clothianidin, thiamethoxam) are selective agonists of insect nicotinic acetylcholine receptors. Although their mammalian toxicity is relatively low, neonicotinoids are persistent and water-soluble, posing long-term risks to aquatic systems [23–26].

- **Organochlorines** (e.g., DDT, lindane, heptachlor), now largely banned or restricted in many regions, are notorious for their high persistence, lipophilicity, and tendency to bioaccumulate in aquatic food webs [27, 28].

- **Herbicides** (e.g., atrazine, glyphosate, paraquat) are designed to disrupt plant-specific processes such as photosynthesis; however, several also interfere with endocrine pathways and oxidative status in aquatic vertebrates [28, 29].

- **Fungicides** (e.g., mancozeb, propiconazole) often affect mitochondrial function and oxidative stress pathways. Their effects on aquatic microbiota and fish are receiving increasing scrutiny [31].

These groups vary in environmental persistence, bioavailability, and toxicity profiles, making their ecotoxicological risk assessments complex and often species-specific [32].

2.2 Environmental pathways and distribution

Once released into the environment, pesticides can enter aquatic systems through several pathways (**Figure 1**):

- **Surface runoff**: Agricultural fields are among the primary sources of pesticide runoff during rainfall events. This is a dominant entry route for pyrethroids and neonicotinoids [2, 33].

- **Spray drift**: Wind dispersion during pesticide application may lead to deposition in nearby water bodies [34].

- **Leaching and subsurface flow**: Certain herbicides and insecticides with high water solubility infiltrate the soil and reach groundwater or connected surface waters [6, 35].

- **Effluents from wastewater treatment plants**: Urban and industrial discharges often contain pesticide residues that escape conventional treatment processes [36, 37].

- **Atmospheric deposition**: Volatile or semi-volatile pesticides, such as endosulfan and lindane, can be transported over long distances and deposited through precipitation [38].

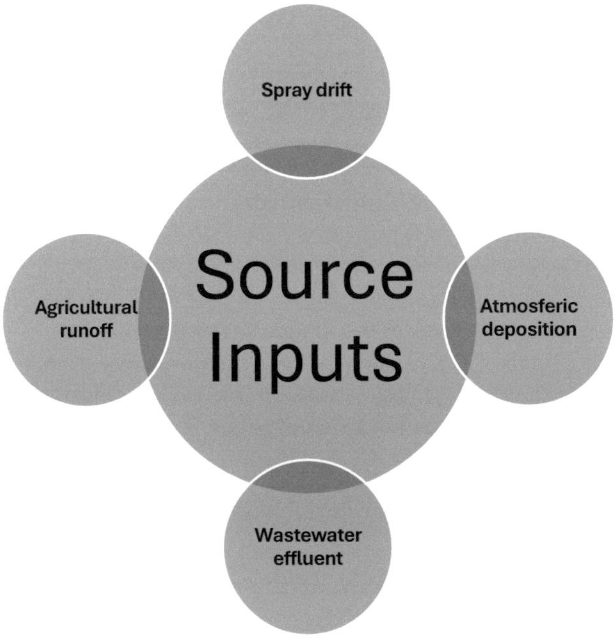

Figure 1.
Sources of pesticide inputs into the environment.

These processes are influenced by climatic, hydrological, and landscape variables, which affect pesticide transport, transformation, and eventual fate [39].

2.3 Fate and persistence in aquatic environments

Pesticide behavior in aquatic ecosystems depends on key physico-chemical parameters:

- **Solubility** determines bioavailability and the potential for leaching.
- **Kow (octanol–water partition coefficient)** indicates lipophilicity and the potential for bioaccumulation.
- **Half-life (DT50)** influences persistence; organochlorines exhibit DT50 values measured in years, whereas organophosphates degrade within days to weeks [40].

Microbial degradation, photolysis, hydrolysis, and adsorption to sediments are principal degradation pathways. However, these processes vary in efficiency across different water bodies [41]. Sorption to particulate matter can reduce bioavailability but also act as a reservoir for slow release, as seen with pyrethroids [42].

2.4 Bioaccumulation and biomagnification potential

Lipophilic pesticides, such as organochlorines and pyrethroids, are characterized by high hydrophobicity, resulting from their chemical structure, which makes them poorly soluble in water. This lipophilicity promotes their entry into biological membranes and fatty tissues once they enter an organism. Therefore, lipophilic pesticides exhibit strong **bioaccumulation** in aquatic species, especially those with high lipid content. Fish, bivalves, and amphibians can accumulate residues in liver, muscle, and fat tissues, sometimes exceeding environmental concentrations by several orders of magnitude [43, 44].

While bioaccumulation refers to the process by which these compounds gradually accumulate in the tissues of living organisms over time, typically through direct absorption from water or through ingestion of contaminated food, biomagnification, on the other hand, describes the increasing concentration of these substances as they move up the food chain, from prey to predator, leading to higher contaminant levels in top predators compared to organisms at lower trophic levels [45].

In food webs, biomagnification occurs when persistent pesticides are transferred and amplified through trophic levels. Top predators, such as piscivorous birds and large fish species, have been shown to accumulate high levels of DDT, leading to reproductive failure and population declines [46].

2.5 Complex mixtures and synergistic effects

Aquatic organisms are rarely exposed to single pesticides. Mixtures can exert additive, antagonistic, or synergistic effects, which complicate toxicological predictions [47, 48]. For example, the combination of atrazine and chlorpyrifos in zebrafish caused more severe neurobehavioral and endocrine effects than either chemical alone [49].

Cumulative exposure to pesticide mixtures is now recognized as a critical factor in ecological risk assessments (Kortenkamp et al., 2007). Sublethal concentrations, once considered environmentally safe, may have significant long-term effects when multiple compounds interact [50].

3. Biotransformation and detoxification mechanisms

3.1 Overview of xenobiotic biotransformation in aquatic organisms

Biotransformation is a critical physiological process by which aquatic organisms modify and detoxify xenobiotics, including pesticides, facilitating their elimination and reducing their toxic potential. This process typically involves a two-phase enzymatic pathway (**Figure 2**):

- **Phase I** (functionalization) includes oxidation, reduction, and hydrolysis reactions, primarily mediated by the cytochrome P450 (CYP) enzyme family (especially the CYP1A, CYP2B, and CYP3A subfamilies).

- **Phase II** (conjugation) involves coupling the Phase I metabolites with hydrophilic endogenous molecules, such as glutathione, glucuronic acid, or sulfate, to enhance solubility and excretion.

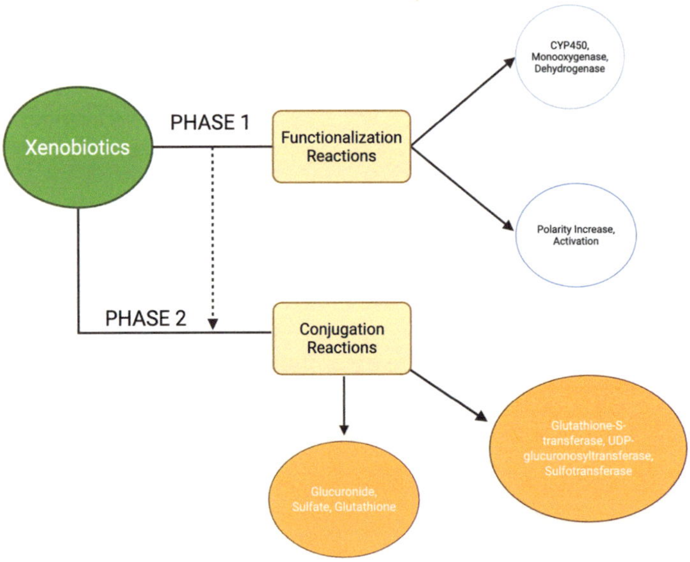

Figure 2.
Xenobiotic metabolism overview.

Biotransformation efficiency varies widely among aquatic species due to interspecies differences in enzymatic systems, developmental stages, nutritional status, and environmental factors such as temperature and oxygen availability [51,52].

3.2 Phase I reactions and the CYP enzyme system

The CYP monooxygenases are a superfamily of heme-thiolate enzymes that catalyze the insertion of oxygen atoms into organic substrates. In fish and invertebrates, CYP1A is often used as a biomarker of exposure to planar aromatic hydrocarbons, pesticides, and other xenobiotics.

CYP1A induction is commonly measured via the following:

- **EROD activity** (ethoxyresorufin-O-deethylase): A sensitive and widely used assay for assessing CYP1A catalytic activity.
- **Gene expression profiling**: quantitative PCR (qPCR) or RNA sequencing (RNA-seq) targeting CYP1A, CYP2B, or CYP3A.

For example, exposure of rainbow trout (*Oncorhynchus mykiss*) to atrazine and chlorpyrifos induced hepatic CYP1A mRNA expression and EROD activity [53]. Similarly, fenvalerate exposure in common carp (*Cyprinus carpio*) led to significant increases in CYP-mediated monooxygenase activity [54].

However, the induction of CYP enzymes is not always protective. In some cases, the bioactivation of pesticides during Phase I metabolism can generate reactive intermediates, such as epoxides or quinones, which cause oxidative stress, mutagenesis, or carcinogenesis [55].

3.3 Phase II detoxification pathways

Phase II reactions conjugate reactive metabolites formed in Phase I with hydrophilic molecules, catalyzed by enzymes such as:

- Glutathione S-transferase (GST)
- UDP-glucuronosyltransferase (UGT)
- Sulfotransferase (SULT)

GST plays a vital role in detoxifying electrophilic pesticide metabolites. Elevated GST activity has been observed in goldfish (*Carassius auratus*) exposed to deltamethrin and in mollusks exposed to triazine herbicides[56,57].

Glucuronidation and sulfation are essential for solubilizing polar metabolites. However, the efficiency of these conjugation reactions may be reduced under stress or in combination with other toxicants, thereby enhancing overall toxicity [58].

3.4 Regulatory mechanisms and xenobiotic sensors

Several transcription factors mediate the cellular response to pesticides and regulate biotransformation enzyme expression:

- **Aryl hydrocarbon receptor**: Activated by polycyclic aromatic compounds and dioxin-like pesticides, it induces CYP1A transcription.
- **Pregnane X receptor and constitutive androstane receptor**: Regulate a broad spectrum of detoxification enzymes.
- **Nuclear factor erythroid 2-related factor 2 (Nrf2)**: Governs antioxidant response elements, modulating GST, SOD, and other detoxification enzymes.

Chlorpyrifos exposure in medaka (*Oryzias latipes*) was shown to activate the Nrf2 pathway and upregulate Phase II enzymes, highlighting a compensatory antioxidant response [59].

3.5 Species-specific and environmental variability

Biotransformation capacity varies among taxa:

- Fish typically possess well-developed CYP and GST systems.
- Mollusks and crustaceans may rely more on esterases and hydrolases.
- Amphibians exhibit ontogenetic variation in metabolic capacity.

Environmental factors such as hypoxia, salinity, and temperature influence enzymatic expression and function. For instance, low oxygen availability suppresses CYP1A activity in zebrafish, altering the detoxification kinetics of pyrethroids [60].

Furthermore, biotransformation pathways may become saturated under high pesticide loads or prolonged exposure, leading to the accumulation of toxic intermediates and oxidative stress [61].

3.6 Role of microbial biotransformation

Microbial communities in fish guts and aquatic sediments also contribute to pesticide degradation. Certain bacterial genera (e.g., *Pseudomonas*, *Sphingomonas*, *Bacillus*) possess plasmid-encoded enzymes capable of hydrolyzing or oxidizing organophosphates and carbamates [62].

Disruption of gut microbiota by pesticides, such as glyphosate and chlorpyrifos, can alter host metabolism, nutrient absorption, and immune function, indirectly affecting the host's detoxification capacity [63].

4. Biomarkers of pesticide toxicity

Biomarkers are measurable indicators of physiological, biochemical, or molecular changes in organisms exposed to environmental contaminants such as pesticides. They serve as early warning tools, often responding to sublethal concentrations and providing insights into mechanisms of toxicity, stress responses, and potential long-term effects. This section explores the most widely applied and emerging biomarker categories in aquatic toxicology, emphasizing their specificity, sensitivity, and ecological relevance (**Figure 3**).

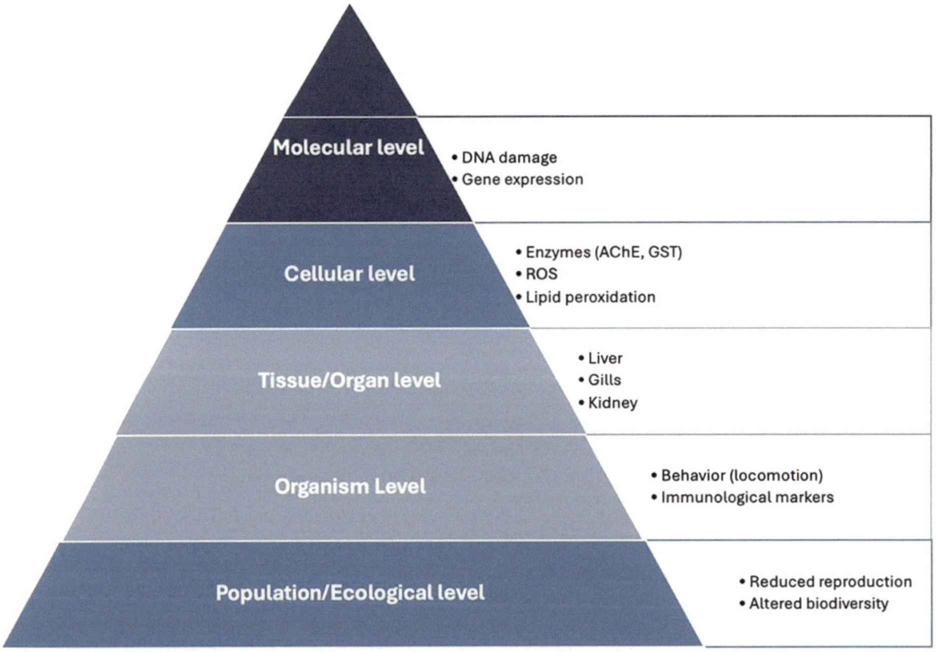

Figure 3.
Scheme of the most widely applied and emerging biomarker categories in aquatic toxicology.

4.1 Biochemical biomarkers

AChE inhibition is the most established biomarker for assessing exposure to organophosphates and carbamates, which inhibit AChE at cholinergic synapses, leading to neurotoxicity and impaired locomotion [15]. Studies in zebrafish (*Danio rerio*), rainbow trout (*Oncorhynchus mykiss*), and tilapia (*Oreochromis niloticus*) have shown significant AChE inhibition upon exposure to diazinon, chlorpyrifos, and malathion [53, 64].

Other enzymatic biomarkers include the following:

- **Lactate dehydrogenase and alanine aminotransferase (ALT)**: Indicators of cellular damage and membrane integrity.

- **Alkaline phosphatase (ALP)**: Linked to hepatobiliary function and tissue repair processes [65].

Substantial changes in plasma ALT, AST, and ALP activities have been reported in fish exposed to pyrethroids and glyphosate-based herbicides Abdel-Tawwab et al., [66], Mahadevan et al., [67].

4.2 Oxidative stress biomarkers

Pesticide exposure often leads to the overproduction of reactive oxygen species, causing oxidative stress and subsequent cellular damage. Biomarkers of oxidative stress include the following:

- **Enzymatic antioxidants**:
    - Superoxide dismutase (SOD)
    - Catalase (CAT)
    - Glutathione peroxidase (GPx)
    - Glutathione reductase (GR)
- **Nonenzymatic indicators**:
    - Glutathione (GSH)/oxidized glutathione (GSSG) ratio
    - Malondialdehyde (MDA): A lipid peroxidation marker
    - 8-hydroxy-2'-deoxyguanosine (8-OHdG): DNA oxidation product.

Oxidative biomarkers are commonly altered in various species following pesticide exposure. For instance, deltamethrin and endosulfan elevated MDA levels and decreased SOD and CAT activity in carp and catfish [68, 69]. Chlorpyrifos exposure in zebrafish also increased oxidative DNA damage, as evidenced by elevated 8-OHdG levels [70].

4.3 Immunological biomarkers

Immune function is sensitive to pesticide-induced modulation. Both immunosuppression and hyperactivation can occur, depending on the dose and compound:

- Lysozyme activity: Reflects innate immune status.
- Respiratory burst activity: Assesses phagocyte function.
- Cytokine expression: IL-1β, TNF-α, and IL-6 are markers of inflammation.
- Complement activity and total immunoglobulin levels.

Subchronic glyphosate exposure suppressed lysozyme activity and the phagocytic index in Nile tilapia, making them more susceptible to opportunistic infections [71]. Immunotoxicity is increasingly associated with gut microbiota disruption, linking systemic inflammation to dysbiosis in fish intestines [72].

4.4 Genetic and genotoxic biomarkers

Genotoxicity biomarkers provide early warnings of DNA damage and are frequently employed in ecotoxicological assessments:

- Comet assay (single-cell gel electrophoresis): Detects DNA strand breaks.
- Micronucleus test: Identifies chromosomal aberrations.
- Sister chromatid exchanges and nuclear abnormalities.

These endpoints have been applied to fish (e.g., *Channa punctatus*, *Oreochromis mossambicus*) exposed to atrazine, pendimethalin, and glyphosate, showing dose-dependent genotoxic responses [73, 74].

4.5 Transcriptomic and proteomic biomarkers

High-throughput technologies now enable comprehensive profiling of gene and protein expression in response to pesticide exposure:

- **Transcriptomics**: Measures mRNA levels using RNA-seq or qPCR arrays. Genes involved in detoxification (CYPs, GST), oxidative stress (SOD, Nrf2), and endocrine pathways (ER, AR) are commonly profiled [12].
- **Proteomics**: Mass spectrometry-based techniques identify differentially expressed proteins involved in apoptosis, mitochondrial dysfunction, and cellular repair [75].

Zebrafish larvae exposed to bifenthrin exhibited altered transcription of genes associated with synaptic signaling and oxidative damage [76].

4.6 Histopathological biomarkers

Histopathological examinations offer visual and structural evidence of tissue-specific toxicity. Common changes include the following:

- Hepatic vacuolization, necrosis, fibrosis
- Gill hyperplasia, lamellar fusion, and clubbing
- Renal tubular degeneration
- Gonadal atresia and testicular degeneration

Exposure to diazinon, endosulfan, and glyphosate has caused significant histological damage to the liver, kidneys, and gills in multiple fish species [53, 77, 78].

4.7 Behavioral biomarkers

Behavioral changes are noninvasive, sensitive endpoints often linked to neurotoxicity:

- Erratic or hypoactivity
- Loss of schooling or predator avoidance
- Impaired feeding or reproduction
- Altered phototaxis or locomotor patterns.

Fenitrothion-exposed zebrafish exhibited impaired escape responses and reduced social interactions, highlighting disruptions in sensorimotor processing [79]. Behavioral assays, including video-tracking and AI-based quantification, are increasingly used for rapid screening.

5. Histopathological and physiological biomarkers

Histopathological and physiological biomarkers offer direct, often organ-specific, insights into the structural and functional impairments caused by pesticide exposure in aquatic organisms. These biomarkers reflect both immediate cytotoxic effects and cumulative long-term tissue damage, enabling correlation with broader organismal and ecological outcomes [80, 81].

5.1 Hepatic alterations

The liver is the primary detoxification organ and a major site for xenobiotic metabolism. Pesticide-induced hepatic injuries are among the most consistently reported histopathological alterations in fish and amphibians.

Common liver pathologies include the following:

- Hepatocellular vacuolization
- Cytoplasmic degeneration
- Necrosis and apoptosis
- Fibrosis and nuclear hypertrophy.

In *Oreochromis mossambicus*, exposure to chlorpyrifos caused extensive cytoplasmic vacuolization, necrotic lesions, and vascular congestion [82]. Similar hepatic damage was documented in zebrafish exposed to glyphosate, including loss of hepatic cord architecture and fatty infiltration [78].

In addition to histological analysis, serum biomarkers such as ALT, AST, and ALP are used to detect liver enzyme leakage due to membrane damage, correlating well with histological findings [67].

5.2 Gill pathology

The gills are critical for gas exchange, osmoregulation, and excretion, making them a sensitive target for waterborne toxicants.

Observed histopathological effects include the following:

- Lamellar fusion and epithelial lifting
- Hyperplasia and hypertrophy
- Clubbing and aneurysms
- Chloride cell proliferation.

Gill lesions have been widely reported in fish exposed to pyrethroids (e.g., deltamethrin), organophosphates (e.g., diazinon), and neonicotinoids (e.g., imidacloprid) [53, 83]. In *Clarias gariepinus*, sublethal exposure to malathion led to extensive epithelial lifting and lamellar necrosis [84, 85].

Gill Na^+/K^+-ATPase activity is often impaired, reflecting compromised osmoregulatory capacity and ionic imbalance under pesticide stress [86].

5.3 Renal damage

The kidney plays a key role in excretion and ion regulation. Histopathological markers of nephrotoxicity include the following:

- Tubular epithelial degeneration
- Glomerular shrinkage and necrosis

- Hyaline droplet accumulation
- Interstitial hemorrhage.

Trichlorfon and cypermethrin have been shown to cause glomerular damage and tubular necrosis in common carp (*Cyprinus carpio*) and African catfish (*Clarias gariepinus*) [87, 88].

In many cases, kidney pathology is accompanied by hematological disruptions, such as decreased hemoglobin levels and erythrocyte counts, signifying systemic toxicity.

5.4 Gonadal and reproductive effects

Pesticides can disrupt endocrine signaling, leading to reproductive toxicity. Histopathological endpoints include the following:

- Follicular atresia and oocyte degeneration
- Spermatogenic arrest and vacuolization of seminiferous tubules
- Interstitial edema and testicular fibrosis.

Atrazine and endosulfan have been linked to altered gonadal structure and impaired gametogenesis in *Danio rerio* and *Rana pipiens* [89, 90]. In male fish, vitellogenin induction and feminization are frequently observed following exposure to estrogenic pesticides.

Reproductive biomarkers are often paired with endocrine indicators, such as hormone levels (testosterone, estradiol) and gene expression of steroidogenic enzymes (e.g., CYP19, StAR).

5.5 Hematological and osmoregulatory biomarkers

Pesticides can alter hematological parameters, reflecting physiological stress, immune suppression, and anemia. Key indicators include the following:

- Hemoglobin concentration (Hb)
- Hematocrit (Ht)
- Red and white blood cell counts (RBC/WBC)
- Mean corpuscular volume and mean corpuscular hemoglobin.

Fenvalerate exposure in *Labeo rohita* caused a significant drop in RBC and Hb levels, indicating hemolysis and anemia [91]. Similarly, glyphosate exposure led to leukopenia and altered leukocyte profiles in tilapia, suggesting immunosuppression [71].

Na^+/K^+-ATPase and carbonic anhydrase activity in gills and kidneys serve as sensitive markers of osmoregulatory function. Imidacloprid and chlorpyrifos have been shown to impair these enzymes, leading to ion imbalance and reduced osmoregulatory performance [92].

5.6 Multiorgan pathology and systemic effects

Environmentally relevant concentrations of pesticides may induce systemic toxicity, affecting multiple organs. Integrated histopathological indices (e.g., Index of Histopathological Alteration) are increasingly used to quantify cumulative damage across organs.

Combined effects, such as liver necrosis, gill degeneration, and renal tubular necrosis, were observed in fish exposed to pesticide mixtures, highlighting the importance of assessing tissue-level damage in combination with biochemical and molecular markers [93].

6. Behavioral and ecological biomarkers

Behavioral and ecological biomarkers provide crucial, often early, insights into how pesticide exposure disrupts organism-level functions and ecosystem dynamics. Behavioral responses are typically rapid, sensitive, and reflect integration across multiple physiological systems, particularly the nervous and endocrine systems. These endpoints are increasingly applied in ecological risk assessments and environmental monitoring.

6.1 Neurobehavioral alterations

Behavioral biomarkers related to neurotoxicity are particularly relevant for pesticides that interfere with synaptic transmission, such as organophosphates, carbamates, and neonicotinoids.

Common behavioral endpoints include the following:

- Impaired locomotion (e.g., hypoactivity or hyperactivity)
- Erratic swimming and loss of equilibrium
- Startle response attenuation
- Altered predator avoidance.

For example, chlorpyrifos exposure in zebrafish larvae significantly reduced swimming activity, responses to light/dark transitions, and escape behavior [94]. Similarly, fenitrothion exposure impaired sensory-motor reflexes and phototactic behavior in early life stages [79].

Neonicotinoids, such as imidacloprid and thiamethoxam, disrupt cholinergic signaling, leading to tremors, paralysis, and behavioral rigidity in aquatic insects and fish [95, 96].

6.2 Feeding and foraging behavior

Feeding impairment is a sensitive indicator of sublethal pesticide toxicity and is associated with metabolic suppression or neurophysiological dysfunction.

Endpoints include the following:

- Reduced food intake
- Prolonged feeding latency
- Disrupted prey detection or handling time.

Pyrethroid exposure in goldfish (*Carassius auratus*) reduces feeding motivation and alters prey capture efficiency, which could compromise growth and survival [97]. In amphibians, sublethal glyphosate levels reduce tadpole foraging success and alter dietary preferences [98].

6.3 Social and reproductive behavior

Pesticides can alter social hierarchies, courtship, and mating behavior by disrupting hormonal and olfactory cues.

Examples include the following:

- Diminished mating calls and courtship displays
- Decreased nest guarding or territorial aggression
- Delayed spawning or altered mating frequency.

Atrazine and endosulfan exposures in frogs and fish were associated with the feminization of males, the loss of secondary sexual traits, and impaired reproductive success [49, 99].

6.4 Cognitive and learning effects

Emerging research has shown that pesticide exposure may impair cognitive abilities in fish, including associative learning, memory retention, and spatial navigation.

Zebrafish exposed to sublethal levels of diazinon exhibited diminished performance in maze-learning tasks and memory recall [100]. These deficits are likely mediated through the disruption of cholinergic and dopaminergic neurotransmission.

6.5 Ecological-level responses

Beyond individual-level behaviors, pesticides can induce community and ecosystem-level changes:

- Decline of sensitive taxa (e.g., Ephemeroptera, Trichoptera)
- Shifts in trophic interactions and predator–prey balance
- Altered species richness and functional diversity.

Mesocosm studies show that pesticide mixtures reduce biodiversity and impair ecosystem functions, such as primary production, decomposition, and nutrient cycling [101, 102].

6.6 Integrative tools and approaches

New technologies enable high-throughput behavioral assessments using automated tracking software, which can quantify endpoints such as velocity, distance moved, turning angles, and inter-individual interactions [103, 104].

- EthoVision, ZebraLab, and DanioVision are popular systems used to evaluate behavior in fish.
- Machine learning algorithms now assist in detecting subtle behavioral changes that may go unnoticed through visual observation [105].

These tools enhance the sensitivity and reproducibility of behavioral biomarker applications and support ecological monitoring programs.

7. Microbiome and omics-based biomarkers

Advances in molecular biology and bioinformatics have enabled the development of highly sensitive and comprehensive biomarker systems. Among these, microbiome composition and omics-based biomarkers (transcriptomics, proteomics, metabolomics, and epigenomics) provide holistic insights into the sublethal and chronic impacts of pesticide exposure on aquatic organisms [106].

7.1 Gut microbiota alterations

The gut microbiome plays essential roles in host digestion, immune modulation, and xenobiotic metabolism. Pesticides can disrupt microbial homeostasis, leading to dysbiosis – a state associated with impaired nutrient absorption, inflammation, and altered immunity [63].

Pesticide-induced microbiota changes have been reported in various aquatic species:

- Carp (*Cyprinus carpio*) exposed to trichlorfon exhibited reduced microbial diversity and significant shifts in the Firmicutes/Bacteroidetes ratio, which correlated with increased intestinal inflammation and permeability [107].
- Zebrafish larvae exposed to glyphosate exhibited a decreased abundance of beneficial genera, such *as Lactobacillus,* and increased levels of opportunistic pathogens, including *Aeromonas* and *Pseudomonas* [108].
- Chlorpyrifos exposure in medaka (*Oryzias latipes*) altered short-chain fatty acid (SCFA)-producing bacterial populations, reducing SCFA availability and impacting gut–brain signaling [109].

Microbiome profiling using 16S rRNA sequencing, metagenomics, and metabolomics can, therefore, act as a powerful diagnostic tool for pesticide toxicity.

7.2 Transcriptomics

Transcriptomic biomarkers provide a genome-wide view of gene expression changes in response to chemical stressors. Techniques such as RNA-seq and qPCR have identified key transcriptional alterations related to oxidative stress, apoptosis, endocrine disruption, and immune responses.

- Exposure to atrazine and chlorpyrifos induced differential expression of detoxification-related genes (e.g., *CYP1A*, *GST*, *UGT*) in *Danio rerio* and *Oncorhynchus mykiss* [110].

- Fenitrothion exposure in zebrafish embryos led to the downregulation of neurodevelopmental genes and the upregulation of inflammatory markers, highlighting neuroimmunotoxic interactions [76].

- Thiacloprid exposure significantly disrupted circadian rhythm-associated genes in medaka, suggesting effects on behavior and hormonal cycles [111].

Pathway enrichment analysis of transcriptomic data enables the mapping of altered gene networks, thereby improving mechanistic understanding.

7.3 Proteomics

Proteomics assesses protein-level alterations, capturing posttranslational modifications and degradation products. Mass spectrometry (LC–MS/MS) and 2D gel electrophoresis are widely used to quantify protein expression and structure.

- Carbamate pesticide exposure in goldfish revealed dysregulation of proteins involved in oxidative phosphorylation, apoptosis, and cytoskeletal structure [75].

- Organophosphate exposure in tilapia led to altered levels of heat shock proteins (HSP70), indicating cellular stress [112].

Proteomic biomarkers are advantageous due to their closer linkage to phenotype than transcript levels, as well as their sensitivity to cumulative stress.

7.4 Metabolomics

Metabolomics involves the comprehensive profiling of endogenous metabolites, offering a snapshot of physiological status.

- Zebrafish exposed to bifenthrin showed disruptions in amino acid and energy metabolism, particularly in pathways involving glutamate and citric acid cycle intermediates [113].

- Atrazine exposure altered hepatic levels of glucose, lactate, and free fatty acids, suggesting metabolic reprogramming [114].

High-resolution NMR and LC–MS are typical platforms used in metabolomic analysis, providing robust biomarker signatures for diagnostic purposes.

7.5 Epigenetic biomarkers

Epigenetic modifications regulate gene expression without altering DNA sequences. These include DNA methylation, histone acetylation, and noncoding RNAs. Pesticides can cause heritable epigenetic changes, which may lead to transgenerational effects.

- Bisphenol A and atrazine exposure in zebrafish resulted in promoter methylation of estrogen receptor genes, which was associated with abnormal sex differentiation [115].

- Malathion exposure triggered miRNA deregulation, affecting immune-related genes in common carp [116].

Epigenetic biomarkers offer insights into the long-term and latent effects of early-life pesticide exposure.

7.6 Integrative multi-omics approaches

Multi-omics strategies combine genomic, transcriptomic, proteomic, metabolomic, and microbiomic data to provide a systems-level understanding of pesticide toxicity. Such integration enables the following:

- Identification of mechanistic pathways

- Discovery of novel biomarkers

- Improved prediction of toxicological outcomes.

For instance, integrated omics analysis in medaka exposed to chlorpyrifos revealed crosstalk among gut microbiome alterations, hepatic lipid metabolism, and neurobehavioral changes [109].

8. Conclusion

Pesticides pose significant risks to aquatic organisms through multiple toxic pathways. Biomarkers provide essential tools for detecting and quantifying sublethal effects of these compounds, ranging from molecular alterations to ecological disturbances. The integration of biochemical, genetic, and behavioral biomarkers enhances environmental monitoring and facilitates early intervention before irreversible damage occurs. Future research should emphasize multi-biomarker approaches and explore novel endpoints, such as transcriptomics and microbiome shifts, to strengthen ecological risk assessments.

Author details

Davide Di Paola[1*], Cristiana Roberta Multisanti[2], Mahdi Banaee[3], Federica Impellitteri[2] and Caterina Faggio[1,4]

1 Department of Chemical, Biological, Pharmaceutical, and Environmental Sciences, University of Messina, Messina, Italy

2 Department of Veterinary Sciences, University of Messina, Messina, Italy

3 Department of Fishery Sciences, Faculty of Natural Resources, Behbahan Khatam Alanbia University of Technology, Behbahan, Iran

4 Department of Eco-sustainable Marine Biotechnology, Stazione Zoologica Anton Dohrn, Naples, Italy

*Address all correspondence to: davide.dipaola@unime.it

References

[1] Schulz R, Bub S, Petschick LL, Stehle S, Wolfram J. Applied pesticide toxicity shifts toward plants and invertebrates, even in GM crops. Science (New York, NY). 2021;**372** (6537):81–84. DOI: 10.1126/science.abe1148

[2] Stehle S, Schulz R. Agricultural insecticides threaten surface waters at the global scale. Proceedings of the National Academy of Sciences of the United States of America. 2015;**112** (18):5750–5755. DOI: 10.1073/pnas.1500232112

[3] Barathinivas A, Ramya S, Neethirajan K, Jayakumararaj R, Pothiraj C, Balaji P, Faggio C. Ecotoxicological effects of pesticides on hematological parameters and oxidative enzymes in freshwater Catfish, Mystus keletius. Sustainability. 2022;**14**:9529. DOI: 10.3390/su14159529

[4] Ramya S, Barathinivas A, Jayakumararaj R, Pothiraj C, Ali D, Piccione G, Multisanti CR, Balaji P. Faggio C. Ecotoxicological insights: Effects of pesticides on ionic metabolism regulation in freshwater catfish, Mystus keletius. Aquatic Toxicology. 2023;**265**:1–7. DOI: 10.1016/j.aquatox.2023.106764

[5] Stara A, Kubec J, Zuskova E, Buric M, Faggio C, Kouba A, Velisek J. Effects ofS-metolachlor and its degradation product metolachlor OA on marbled crayfish (Procambarus virginalis). Chemosphere. 2019;**224**:616–625. DOI: 10.1016/j.chemosphere.2019.02.187

[6] Pérez-Lucas G, Vela N, El Aatik A, Navarro S. Environmental risk of groundwater pollution by pesticide leaching through the soil profile Pesticides. Use and Misuse and Their Impact in the Environment. 2019;**1**:267–298. DOI: 10.5772/intechopen.82418

[7] Wang L, Zheng M, Xu H, Hua Y, Liu A, Li Y, Fang L, Chen X. Fate and ecological risks of current-use pesticides in seawater and sediment of the Yellow Sea and East China Sea. Environmental Research. 2022;**207**:112673. DOI: 10.1016/j.envres.2021.112673

[8] Impellitteri F, Curpăn AS, Plăvan G, Ciobica A, Faggio C. Hemocytes: A useful tool for assessing the toxicity of microplastics, heavy metals, and pesticides on aquatic invertebrates. International Journal of Environmental Research and Public Health. 2022;**19** (24):1–15. DOI: 10.3390/ijerph192416830

[9] Banaee M, Di Paola D, Cuzzocrea S, Cordaro M, Faggio C. Biomarkers in Aquatic Ecotoxicology: Understanding the Effects of Xenobiotics on the Health of Aquatic Organisms. In: Biochemistry. IntechOpen; 2024. DOI: 10.5772/intechopen.1006063

[10] van der Oost R, Beyer J, Vermeulen NP. Fish bioaccumulation and biomarkers in environmental risk assessment: a review. Environmental Toxicology and Pharmacology. 2003;**13** (2):57–149. DOI: 10.1016/s1382-6689(02)00126-6

[11] Ankley GT, Bennett RS, Erickson RJ, Hoff DJ, Hornung MW, Johnson RD, Mount DR, Nichols JW, Russom CL, Schmieder PK, Serrrano JA, Tietge JE, Villeneuve DL. Adverse outcome pathways: a conceptual framework to support ecotoxicology research and risk

assessment. Environmental Toxicology and Chemistry. 2010;**29**(3):730–41. DOI: 10.1002/etc.34

[12] Sumpter JP. Endocrine disrupters in the aquatic environment: An overview. Acta Hydrochimica Et Hydrobiologica. 2005;**33**(1):9–16. DOI: 10.1002/aheh.200400555

[13] Köhler HR, Triebskorn R. Wildlife ecotoxicology of pesticides: Can we track effects to the population level and beyond?. Science. 2013;**341**(6147):759–765. DOI: 10.1126/science.1237591

[14] Sula E, Aliko V, Marku E, Nuro A, Faggio C. Evaluation of kidney histopathological alterations in Crucian Carp, Carassius carassius, from a pesticide and PCB- contaminated freshwater ecosystem, using light microscopy and organ index mathematical model. International Journal of Aquatic Biology. 2020;**8**(3):154–165. https://ij-aquaticbiology.com/index.php/ijab/article/view/829

[15] Lionetto MG, Caricato R, Calisi A, Giordano ME, Schettino T. Acetylcholinesterase as a biomarker in environmental and occupational medicine: new insights and future perspectives. BioMed Research International. 2013;**2013**:321213. DOI: 10.1155/2013/321213

[16] Tierney KB, Ross PS, Jarrard HE, Delaney KR, Kennedy CJ. Changes in juvenile coho salmon electro-olfactogram during and after short-term exposure to current-use pesticides. Environmental Toxicology and Chemistry. 2006;**25**(10):2809-17. DOI: 10.1897/05-629r1.1

[17] Banaee M, Impellitteri F, Multisanti CR, Sureda A, Arfuso F, Piccione G, Faggio C. Evaluating Silymarin Extract as a Potent Antioxidant Supplement in Diazinon-Exposed Rainbow Trout: Oxidative Stress and Biochemical Parameter Analysis. Toxics. 2023;11737. DOI: 10.3390/toxics110907

[18] Saha S, Chukwuka AV, Mukherjee D, Patnaik L, Nayak S, Dhara K, Saha NC, Faggio C. Chronic Effects of Diazinon® Exposures Using Integrated Biomarker Responses in Freshwater Walking Catfish. Clariasbatrachus. Applied Sciences. 2021;**11**:10902. DOI: 10.3390/app112210902

[19] Gupta Ramesh. Toxicology of Organophosphate & Carbamate Compounds. 2006. DOI: 10.1016/B978-0-12-088523-7.X5000-5

[20] Amweg EL, Weston DP, Ureda NM. Use and toxicity of pyrethroid pesticides in the Central Valley, California, USA. Environmental toxicology and chemistry. 2005;**24**(4):966–972. DOI: 10.1897/04-146r1.1

[21] Bradbury SP, Coats JR. Comparative toxicology of the pyrethroid insecticides. Reviews of environmental contamination and toxicology. 1989;**108**:133–177. DOI: 10.1007/978-1-4613-8850-0_4

[22] Jindal R, Sharma R, Kaur P, Kaur S, Multisanti CR, Faggio C. Mitigation of haemato-genotoxic and stress response effects in Cyprinus carpio via silymarin dietary supplementation following deltamethrin exposure. Heliyon. 2024;**10**(7):e28419. DOI: 10.1016/j.heliyon.2024.e28419

[23] Gibbons D, Morrissey C, Mineau P. A review of the direct and indirect effects of neonicotinoids and fipronil on vertebrate wildlife. Environmental science and pollution research

international. 2015;**22**(1):103–118. DOI: 10.1007/s11356-014-3180-5

[24] Morrissey CA, Mineau P, Devries JH, Sanchez-Bayo F, Liess M, Cavallaro MC, Liber K. Neonicotinoid contamination of global surface waters and associated risk to aquatic invertebrates: a review. Environment international. 2015;**74**:291–303. DOI: 10.1016/j.envint.2014.10.024

[25] Mukherjee D, Saha S, Chukwuka AV, Ghosh B, Dhara K, Saha NC, Pal P, Faggio C. Antioxidant enzyme activity and pathophysiological responses in the freshwater walking catfish, Clarias batrachus Linn under sub- chronic and chronic exposures to the neonicotinoid, Thiamethoxam®. Science of the Total Environment. 2022;**836**:1–16. DOI: 10.1016/j.scitotenv.2022.155716

[26] Mandal AH, Sadhu A, Ghosh S, Saha NC, Mossotto C, Pastorino P, Saha S, Faggio C. Evaluating the impact of neonicotinoids on aquatic non-target species: A comprehensive review. Environmental Toxicology and Pharmacology. 2025;**113**:1–15. DOI: 10.1016/j.etap.2024.104606

[27] Turusov V, Rakitsky V, Tomatis L. Dichlorodiphenyltrichloroethane (DDT): ubiquity, persistence, and risks. Environmental Health Perspectives. 2002;**110**(2):125–128. DOI: 10.1289/ehp.02110125

[28] Ritter L, Solomon K, Forget J, Stemeroff M, O'leary C. A Review of Selected Persistent Organic Pollutants. International Programme on Chemical Safety (IPCS). PCS/95.39. World Health Organization, Geneva. 1995;**65**:66

[29] Glusczak L, dos Santos Miron D, Crestani M, Braga da Fonseca M, Pedron A, de F, Duarte MF, Vieira VLP. Effect of glyphosate herbicide on acetylcholinesterase activity and metabolic and hematological parameters in piava (Leporinus obtusidens). Ecotoxicology and Environmental Safety. 2006;**65**(2):237–241. DOI: 10.1016/j.ecoenv.2005.07.017

[30] Solomon KR, Baker DB, Richards RP, Dixon KR, Klaine SJ, La Point TW, Kendall RJ, Weisskopf CP, Giddings JM, Giesy JP, Hall LW, Williams WM. Ecological risk assessment of atrazine in North American surface waters. Environmental Toxicology and Chemistry. 1996;**15**:31–76. DOI: 10.1002/etc.5620150105

[31] Elskus Adria A. Toxicity, sublethal effects, and potential modes of action of select fungicides on freshwater fish and invertebrates. No. 2012–1213. US Geological Survey. 2012. DOI: 10.3133/ofr20121213

[32] Liess M, Beketov M. Traits and stress: Keys to identify community effects of low levels of toxicants in test systems. Ecotoxicology. 2011;**20**(6):1328–1340. DOI: 10.1007/s10646-011-0689-y

[33] Kreuger J. Pesticides in stream water within an agricultural catchment in southern Sweden. Science of the Total Environment. 1998;**216**(3):227–251. DOI: 10.1016/s0048-9697(98)00155-7

[34] Adriaanse P, Boivin A, Klein M, Jarvis N, Stemmer M, Fait G, Egsmose M. EFSA (European Food Safety Authority). Scientific report of EFSA on the 'repair action' of the FOCUS surface water scenarios. EFSA Journal. 2020;**18**(6):e06119. DOI: 10.2903/j.efsa.2020.6119

[35] Bayrak M, et al. Physiological and histopathological response of non-target organisms to emamectin benzoate: An effective innovative insecticide or a new ecosystem health hazard?. Comparative Biochemistry and Physiology Part C: Toxicology and Pharmacology. 2025;**296**:110241

[36] Loos R, Carvalho R, António DC, Comero S, Locoro G, Tavazzi S, Paracchini B, Ghiani M, Lettieri T, Blaha L, Jarosova B, Voorspoels S, Servaes K, Haglund P, Fick J, Lindberg RH, Schwesig D, Gawlik BM. EU-wide monitoring survey on emerging polar organic contaminants in wastewater treatment plant effluents. Water research. 2013;**47**(17):6475–6487. DOI: 10.1016/j.watres.2013.08.024

[37] Gaioto FC, Matheus M, de Souza Chaves BM, Campos JC, Barra TA, de Almeida Azevedo D, Bassin JP, de Carvalho Dezotti MW. Treatment and reuse of a pesticide-containing wastewater by a combination of physicochemical, biological and membrane processes. 2023. DOI: 10.21203/rs.3.rs-2804636/v1

[38] Bidleman TF. Atmospheric Transport and Air-Surface Exchange of Pesticides. Water, Air, and Soil Pollution. 1999;**115**:115–166. DOI: 10.1023/A:1005249305.515

[39] Gilliom RJ. Pesticides in US. streams and groundwater. Environmental Science and Technology. 2007;**41**(10):3408–3414. DOI: 10.1021/es072531u

[40] Howard PH. Handbook of Environmental Fate and Exposure Data for Organic Chemicals. UK: CRC Press; 1991. 712 p. DOI: 10.1201/9780203719305

[41] Cerejeira MJ, et al. Pesticides in Portuguese surface and ground waters. Water Research. 2003;**37**:1055–1063

[42] Weston DP, You J, Lydy MJ. Distribution and Toxicity of Sediment-Associated Pesticides in Agriculture-Dominated Water Bodies of California's Central Valley. Environmental Science and Technology. 2004;**38**(10):2752–2759. DOI: 10.1021/es0352193

[43] Miniero R, Guandalini E, Dellatte E, Iacovella N, Abate V, De Luca S, Iamiceli AL, di Domenico A, De Felip E. Persistent organic pollutants (POPs) in fish collected from the urban tract of the river Tiber in Rome (Italy). Annali dell'Istituto superiore di sanita. 2011;**47**(3):310–315. DOI: 10.4415/ANN_11_03_12

[44] Gui D, Yu R, He X, Tu Q, Chen L, Wu Y. Bioaccumulation and biomagnification of persistent organic pollutants in Indo-Pacific humpback dolphins (Sousa chinensis) from the Pearl River Estuary, China. Chemosphere. 2014;**114**:106–113. DOI: 10.1016/j.chemosphere.2014.04.028

[45] Aliko V, et al. Get rid of marine pollution: Bioremediation an innovative, attractive, and successful cleaning strategy. Sustainability. 2022;**14**(18):1–17. DOI: 10.3390/su141811784

[46] Peakall DB, Lincer JL. Do PCBs cause eggshell thinning?. Environmental Pollution. 1996;**91**(1):127–129. DOI: 10.1016/0269-7491(95)00012-g

[47] Backhaus T, Faust M. Predictive environmental risk assessment of chemical mixtures: A conceptual framework. Environmental Science and Technology. 2012;**46**(5):2564–2573. DOI: 10.1021/es2034125

[48] Banaee M, Zeidi A, Gholamhosseini A, Shakeri R, Faggio C, Multisanti CR. Potential synergistic effects of microplastics and zinc oxide nanoparticles: Biochemical and physiological analysis on Astacus leptodactylus. Ecotoxicology. 2025;1–10. DOI: 10.1007/s10646-025-0 2872-1

[49] Levin ED, Chrysanthis E, Yacisin K, Linney E. Chlorpyrifos exposure of developing zebrafish: effects on survival and long-term effects on response latency and spatial discrimination. Neurotoxicology and Teratology. 2003;**25**(1):51–57. DOI: 10.1016/S0892-0362(02)00322-7

[50] Cedergreen N. Quantifying synergy: A systematic review of mixture toxicity studies within environmental toxicology. PLOS ONE. 2014;**9**(5): e96580. DOI: 10.1371/journal. pone.0096580

[51] Viganó L, Arillo A, Bagnasco M, Melodia F, Galassi S. Xenobiotic metabolizing enzymes in rainbow trout exposed to river water. Comparative biochemistry and physiology. C, Comparative pharmacology and toxicology. 1991;**100**(1-2):169–172. DOI: 10.1016/0742-8413(91)90147-

[52] Goksøyr A. Use of cytochrome P450 1A (CYP1A) in fish as a biomarker of aquatic pollution. In: Degen GH, Seiler JP, Bentley P, editors. Toxicology in Transition. Archives of Toxicology. Vol. **17**. Berlin, Heidelberg: Springer; 1995. DOI: 10.1007/978-3-642-79451-3_7

[53] Julieta S. De Anna, Juan M. Castro, Luis Arias Darraz, Federico D. Elías, Juan G. Cárcamo, Carlos M. Luquet, Exposure to hydrocarbons and chlorpyrifos alters the expression of nuclear receptors and antioxidant, detoxifying, and immune response proteins in the liver of the rainbow trout. Oncorhynchus mykiss, Ecotoxicology and Environmental Safety. 2021;**208**:111394. DOI: 10.1016/j. ecoenv.2020.111394

[54] Xing H, Zhang Z, Yao H, Liu T, Wang L, Xu S, Li S. Effects of atrazine and chlorpyrifos on cytochrome P450 in common carp liver. Chemosphere. 2014;**104**:244–250. DOI: 10.1016/j. chemosphere.2014.01.002

[55] Livingstone DR. Contaminant-stimulated reactive oxygen species production and oxidative damage in aquatic organisms. Marine pollution bulletin. 2001;**42**(8):656–666. DOI: 10. 1016/s0025-326x(01)00060-1

[56] Milan M, Dalla Rovere G, Smits M, Ferraresso S, Pastore P, Marin M. G, Bogialli S, Patarnello T, Bargelloni L, Matozzo V. Ecotoxicological effects of the herbicide glyphosate in non-target aquatic species: Transcriptional responses in the mussel Mytilus galloprovincialis. Environmental Pollution (Barking, Essex: 1987). 2018;**237**. DOI: 10.5//2/intecho pen.101263/442–451

[57] Sayeed I, Parvez S, Pandey S, Bin-Hafeez B, Haque R, Raisuddin S. Oxidative stress biomarkers of exposure to deltamethrin in freshwater fish. Channa punctatus Bloch, Ecotoxicology and Environmental Safety. 2003;**56** (2):295–301. DOI: 10.1016/S0147-6513(03)00009-5

[58] Shimada T. Xenobiotic-Metabolizing Enzymes Involved in Activation and Detoxification of Carcinogenic Polycyclic Aromatic Hydrocarbons. Drug metabolism and pharmacokinetics. 2006;**21**:257–76. DOI: 10.2133/dmpk.21.257

[59] Jeon HJ, Lee YH, Mo, HH, Kim MJ, Al-Wabel MI, Kim Y, Cho K, Kim TW, Ok YS, Lee SE. Chlorpyrifos-induced biomarkers in Japanese medaka (Oryzias latipes). Environmental science and pollution research international. 2015;**23**. DOI: 10.1007/s11356-015-4598-0

[60] Rahman MS, Thomas P. Effects of hypoxia exposure on hepatic cytochrome P450 1A (CYP1A) expression in Atlantic croaker: molecular mechanisms of CYP1A down-regulation. PLoS One. 2012;7(7): e40825. DOI: 10.1371/journal.pone.0040825

[61] Khatib I, Rychter P, Falfushynska H. Pesticide Pollution: Detrimental Outcomes and Possible Mechanisms of Fish Exposure to Common Organophosphates and Triazines. Journal of Xenobiotics. 2022;**12**(3):236–265. DOI: 10.3390/jox12030018

[62] Pujar NK, Premakshi HG, Ganeshkar MP, Kamanavalli CM. Biodegradation of Pesticides Used in Agriculture by Soil Microorganisms. 2022. DOI: 10.1007/978-981-16-4574-7_11

[63] Jin Y, Wu S, Zeng Z, Fu Z. Effects of environmental pollutants on gut microbiota. Environmental Pollution (Barking, Essex: 1987). 2017;**222**:1–9. DOI: 10.1016/j.envpol.2016.11.045

[64] Heba S. Hamed. Impact of a Short – Term Malathion Exposure of Nile Tilapia, (Oreochromis niloticus): The Protective Role of Selenium. International Journal of Environmental Monitoring and Analysis. 2015;**3**(5-1):30–37. DOI: 10.11648/j.ijema.s.2015030501.15

[65] Pereira Trídico C, Ferreira Rodrigues AC, Nogueira L, da Silva D. C, Benedito Moreira A, de Almeida EA. Biochemical biomarkers in Oreochromis niloticus exposed to mixtures of benzo[a]pyrene and diazinon. Ecotoxicology and environmental safety. 2010;**73**(5):858–863. DOI: 10.1016/j.ecoenv.2010.01.016

[66] Abdel-Tawwab M. Effects of dietary protein levels and rearing density on growth performance and stress response of Nile tilapia. International Aquatic Research. 2012;**4**(1):3. DOI: 10.1186/2008-6970-4-3

[67] Maha Samanta P, Pal S, Mukherjee AK, Senapati T, Jung J, Ghosh AR. Assessment of adverse impacts of glyphosate- based herbicide, Excel Mera 71 by integrating multi-level biomarker responses in fishes. International Journal of Environmental Science and Technology. 2018;**16**:6291–6300. DOI: 10.1007/s13762-018-2013-3

[68] Formicki G, Goc Z, Bojarski B, Witeska M. Oxidative Stress and Neurotoxicity Biomarkers in Fish Toxicology. Antioxidants (Basel). 2025;**14**(8):939. DOI: 10.3390/antiox14080939

[69] Ural MŞ. Chlorpyrifos- induced changes in oxidant/antioxidant status and haematological parameters of Cyprinus carpio carpio: ameliorative effect of lycopene. Chemosphere. 2013;**90**(7):2059–2064. DOI: 10.1016/j.chemosphere.2012.12.006

[70] Sulihah NT, Putri ER. Positive Correlation Between Pesticide Combination Exposure Risk Stratification And 8-Ohdg Levels As A Dna Damage Biomarker In Farmers.

Buletin Veteriner Udayana. 2025;1132–1137. DOI: 10.24843/bulvet.2025

[71] Ayoola SO. Histopathological Effects of Glyphosate on Juvenile African Catfish (Clarias gariepinus). American-Eurasian Journal of Agricultural and Environmental Sciences. 2007;**4**(3):362–367

[72] Ali A, AlHussaini KI. Pesticides: Unintended Impact on the Hidden World of Gut Microbiota. Metabolites. 2024;**14**(3):155. DOI: 10.3390/metabo14030155

[73] Cavas T. In vivo genotoxicity evaluation of atrazine and atrazine-based herbicide on fish Carassius auratus using the micronucleus test and the comet assay. Food and chemical toxicology: an international journal published for the British Industrial Biological Research Association. 2011;**49**(6):1431–1435. DOI: 10.1016/j.fct.2011.03.038

[74] Gupta P, Verma SK. Evaluation of genotoxicity induced by herbicide pendimethalin in freshwater fish Clarias batrachus (linn.) and possible role of oxidative stress in induced DNA damage. Drug and chemical toxicology. 2022;**45**(2):750–759. DOI: 10.1080/01480545.2020.1774603

[75] Biales AD, Bencic DC, Flick RL, Blocksom KA, Lazorchak JM, Lattier DL. Proteomic analysis of a model fish species exposed to individual pesticides and a binary mixture. Aquatic Toxicology (Amsterdam, Netherlands). 2011;**101**(1):196–206. DOI: 10.1016/j.aquatox.2010.09.019

[76] Frank DF, Miller GW, Harvey DJ, Brander SM, Geist J, Connon RE, Lein PJ. Bifenthrin causes transcriptomic alterations in mTOR and ryanodine receptor-dependent signaling and delayed hyperactivity in developing zebrafish (Danio rerio). Aquatic Toxicology. 2018;**200**:50–61. DOI: 10.1016/j.aquatox.2018.04.003

[77] Awad SF, Al-Mahmood SS. Effects of glyphosate in common carp: Histopathological and immunohistochemical study. Iraqi Journal of Veterinary Sciences. 2023;**37**(3):659–666. DOI: 10.33899/ijvs.2023.136627.2601

[78] Al-Otaibi AM, Al-Balawi HF, Ahmad Z, Suliman EM. Toxicity bioassay and sub-lethal effects of diazinon on blood profile and histology of liver, gills and kidney of catfish. Clarias gariepinus. Brazilian Journal of Biology. 2019;**79**:326–336. DOI: 10.1590/1519-6984.185408

[79] Faria M, Prats E, Ramírez JR, Bellot M, Bedrossiantz J, Pagano M, Valls A, Gomez-Canela C, Porta JM, Mestres J, Garcia-Reyero N. Androgenic activation, impairment of the monoaminergic system and altered behavior in zebrafish larvae exposed to environmental concentrations of fenitrothion. Science of The Total Environment. 2021;**775**:145671. DOI: 10.1016/j.scitotenv.2021.145671

[80] Chromcova L, Blahova J, Zivna D, Plhalova L, Casuscelli F, Di Tocco L, Prokes M, Faggio C, Tichy F, Svobodova Z. NeemAzal T/S - toxicity to early-life stages of common carp (Cyprinus carpio L.). Veterinární medicína. 2015;**60**:23–30. DOI: 10.17221/7922-VETMED

[81] Kannan M, Bojan N, Swaminathan J, Zicarelli G, Hemalatha D, Zhang Y, Ramesh M, Faggio C. Nanopesticides in agricultural pest management and their

environmental risks: A review. International Journal of Environmental Science and Technology. 2023;**20**:10507–10532. DOI: 10.1007/s13762-023-04795-y

[82] Hossain MA, Sutradhar L, Sarker TR, Saha S, Iqbal MM. Toxic effects of chlorpyrifos on the growth, hematology, and different organs histopathology of Nile tilapia. Oreochromis niloticus. Saudi Journal of Biological Sciences. 2022;**29**(7):103316. DOI: 10.1016/j.sjbs.2022.103316

[83] Majumder R, Kaviraj A. Histopathological Alterations of Gills, Liver and Kidney of Freshwater Fish, Oreochromis niloticus, Exposed to Cypermethrin. Journal of Aquatic Biology and Fisheries. 2022;**10**:1–5. DOI: 10.5//2/intechopen.101263/

[84] Isiyaku Mohammed A, Yusuf T, Laura A, Maimuna AA. Acute Toxicity and Histological Changes in Gills and Liver of Juvenile of Clarias gariepinus expose to Organophosphate Pesticide (DD-Force). International Journal of Oceanography and Aquaculture. 2024;**8**:1–11. DOI: 10.23880/ijoac-16000290

[85] Sula E, Aliko V, Barceló D, Faggio C. Combined effects of moderate hypoxia, pesticide and PCBs upon crucian carp fish, Carassius carassius, from a freshwater lake-in situ ecophysiological approach. Aquatic Toxicology. 2020;**228**:105644. DOI: 10. 1016/j.aquatox.2020.105644

[86] Lacy B, Rahman MS, Rahman MS. Potential mechanisms of Na+/K +-ATPase attenuation by heat and pesticides co-exposure in goldfish: role of cellular apoptosis, oxidative/nitrative stress, and antioxidants in gills. Environmental Science and Pollution Research. 2022;**29**:57376–57394. DOI: 10.1007/s11356-022-19779-739

[87] Burch E, Hussein MA, Zaki M, Kamal LT, Zaki G, Shoeib T, Dawood M, Sewilam H, Abdelnaser A. Assessing the Effects of Pesticides on Aquacultured Fish and Ecosystems: A Comprehensive Environmental Health Review. Fishes. 2025;**10**:223. DOI: 10.3390/fishes10050223

[88] Pal S, Kokushi E, Koyama J, Uno S, Ghosh AR. Histopathological alterations in gill, liver and kidney of common carp exposed to chlorpyrifos. Journal of environmental science and health. Part B, Pesticides, Food Contaminants, and Agricultural Wastes. 2012;**47**(3):180–195. DOI: 10.1080/03601234.2012.632285

[89] Solomon KR, Carr JA, Du Preez LH, Kendall RJ, Smith EE, Van Der Kraak GJ. Effects of Atrazine on Fish, Amphibians, and Aquatic Reptiles: A Critical Review. Critical reviews in toxicology. 2008;**38**:721–72. DOI: 10.10 80/10408440802116496

[90] Hayes TB, Collins A, Lee M, Mendoza M, Noriega N, Stuart AA, Vonk A. Hermaphroditic, demasculinized frogs after exposure to the herbicide atrazine at low ecologically relevant doses. Proceedings of the National Academy of Sciences of the United States of America. 2002;**99** (8):5476–5480. DOI: 10.1073/pnas.082121499

[91] Jasmin J, Rahman MM, Rahman MM. Haematological Changes in Labeo rohita (H.) due to Exposure of Pesticides, Difenoconazole and Thiamethoxam. International Journal of Contemporary Research and Review. 2018. DOI: 10.15520/ijcrr/2018/9/01/390

[92] Alex Z, Perry SF. Physiology and aquaculture: A review of ion and acid-base regulation by the gills of fishes. Fish and Fisheries. 2022;**23**. DOI: 10.1111/faf.12659

[93] Abdallah SM, Muhammed RE, Mohamed RE, El Daous H, Saleh DM, Ghorab MA, Chen S, El-Sayyad GS. Assessment of biochemical biomarkers and environmental stress indicators in some freshwater fish. Environmental geochemistry and health. 2024;**46**(11):464. DOI: 10.1007/s10653-024-02226-6

[94] Levin ED, Chrysanthis E, Yacisin K, Linney E. Chlorpyrifos exposure of developing zebrafish: effects on survival and long-term effects on response latency and spatial discrimination. Neurotoxicology and teratology. 2003;**25**(1):51–57. DOI: 10.1016/s0892-0362(02)00322-7

[95] Rico A, Arenas-Sánchez A, Pasqualini J, García-Astillero A, Cherta L, Nozal L, Vighi M. Effects of imidacloprid and a neonicotinoid mixture on aquatic invertebrate communities under Mediterranean conditions. Aquatic toxicology (Amsterdam, Netherlands). 2018;**204**:130–143. DOI: 10.1016/j.aquatox.2018.09.004

[96] Chung KT, Chen LW, Tseng HW, Wu CH. Neonicotinoid Imidacloprid Affects the Social Behavior of Adult Zebrafish by Damaging Telencephalon Neurons through Oxidation Stress, Inflammation, and Apoptosis. Life (Basel). 2023;**13**(6):1418. DOI: 10.3390/life13061418

[97] Shaluei F, Hedayati A, Kolangi H, Jahanbakhshi A, Baghfalaki M. Evaluation of the Acute Toxicity of Cypermethrin and its Effect on Behavioral Responses of Caspian Roach (Rutilus rutilus caspicus) and Silver Carp (Hypophthalmicthys molitrix). Global Veterinaria. 2012;**9**(2). DOI: 10.5829/idosi.gv.2012.9.2.63235

[98] Relyea RA. The Impact Of Insecticides And Herbicides On The Biodiversity And Productivity Of Aquatic Communities. Ecological Applications. 2005;**15**:618–627. DOI: 10.1890/03-5342

[99] Hayes TB, Khoury V, Narayan A, Nazir M, Park A, Brown T, Adame L, Chan E, Buchholz D, Stueve T, Gallipeau S. Atrazine induces complete feminization and chemical castration in male African clawed frogs (Xenopus laevis). Proceedings of the National Academy of Sciences of the United States of America. 2010;**107**(10):4612–4617. DOI: 10.1073/pnas.0909519107

[100] Boyda J, Hawkey AB, Holloway ZR, Trevisan R, Di Giulio RT, Levin ED. The organophosphate insecticide diazinon and aging: Neurobehavioral and mitochondrial effects in zebrafish exposed as embryos or during aging. Neurotoxicology and Teratology. 2021;**87**:107011. DOI: 10.1016/j.ntt.2021.107011

[101] Beketov MA, Kefford BJ, Schäfer RB, Liess M. Pesticides reduce regional biodiversity of stream invertebrates. Proceedings of the National Academy of Sciences of the United States of America. 2013;**110**(27):11039–11043. DOI: 10.1073/pnas.1305618110

[102] Schäfer RB, Caquet T, Siimes K, Mueller R, Lagadic L, Liess M. Effects of pesticides on community structure and ecosystem functions in agricultural streams of three biogeographical regions in Europe.

Science of the Total Environment. 2007;**382**(2-3):272–285. DOI: 10.1016/j.scitotenv.2007.04.040

[103] Stevanović M, Tagkalidou N, Multisanti CR, Pujol S, Aljabasini O, Prats E, Faggio C, Porta JM, Barata C, Raldúa D. Zebra_K, a kinematic analysis automated platform for assessing sensitivity, habituation and prepulse inhibition of the acoustic startle response in adult zebrafish. Science of the Total Environment. 2025;**958**:178028. DOI: 10.1016/j.scitotenv.2024.178028

[104] Tagkalidou N, Multisanti CR, Bleda MJ, Bedrossiantz J, Prats E, Faggio C, Barata C, Raldúa D. Analyzing the Effects of Age, Time of Day, and Experiment on the Basal Locomotor Activity and Light-Off Visual Motor Response Assays in Zebrafish Larvae. Toxics. 2024;**12**(5):349. DOI: 10.3390/toxics12050349

[105] Peng Y, Hiro T, Masataka M, Motoyuki I. Zebrafish behavior feature recognition using three- dimensional tracking and machine learning. Scientific Reports. 2021;**11**:13492. DOI: 10.1038/s41598-021-92854-0

[106] Burgos-Aceves MA, Banaee M, Vazzana I, Betancourt-Lozano M, González-Mille DJ, Aliko V, Faggio C, Ilizaliturri-Hernández CA. Effect of emerging pollutants on the gut microbiota of freshwater animals: Focusing on microplastics and pesticides. Science of the Total Environment. 2024;**948**:174809. DOI: 10.1016/j.scitotenv.2024.174809

[107] Wang X, Liu P, Yang L, Zheng W, Zhu L, Kong X. Integrated analysis of trichlorfon-induced intestinal toxicity in Micropterus salmoides: Barrier Impairment, microbiome disruption, and metabolic reprogramming. Fish and Shellfish Immunology. 2026;**168**:110988. DOI: 10.1016/j.fsi.2025.110988

[108] Janssens L, Van de Maele M, Delnat V, Theys C, Mukherjee S, De Meester Robby Stoks L. Evolution of pesticide tolerance and associated changes in the microbiome in the water flea Daphnia magna, Ecotoxicology and Environmental Safety. 2022;**240**:113697. DOI: 10.1016/j.ecoenv.2022.113697

[109] Qiu X, Wu Y, Liu Y, Zhuo M, Chen K, Takai Y, Kang I, Shimasaki Y, Oshima Y. Chronic exposure to chlorpyrifos disturbs the locomotor behavior, vocal traits, and brain transcriptional profile of Japanese medaka (Oryzias latipes). Chemico-Biological Interactions. 2026;**432**:112027. DOI: 10.1016/j.cbi.2026.112027

[110] Xing H, Wang Z, Gao X, Chen D, Wang L, Li S, Xu S. Atrazine and chlorpyrifos exposure induces liver autophagic response in common carp. Ecotoxicology and environmental safety. 2015;**113**:52–58. DOI: 10.1016/j.ecoenv.2014.11.027

[111] Zheng X, Shi H, Li B, Li J, Liu Y, Liang J, Chen L, Fent K, Zhang K, Wang H, Dai J, Zhao Y. Identification of toxic substances considering circadian disruption: an integrated approach combining reduced transcriptomic analysis, behavioral monitoring, and time-course gene expression profiling. Environment international. 2026;**208**:110111. DOI: 10.1016/j.envint.2026.110111

[112] Biales AD, Bencic DC, Flick RL, Blocksom KA, Lazorchak JM, Lattier DL. Proteomic analysis of a model fish

species exposed to individual pesticides and a binary mixture. Aquatic toxicology (Amsterdam, Netherlands). 2011;**101**(1):196–206. DOI: 10.1016/j.aquatox.2010.09.019

[113] Bertotto LB, Richards J, Gan J, Volz DC, Schlenk D. Effects of bifenthrin exposure on the estrogenic and dopaminergic pathways in zebrafish embryos and juveniles. Environmental toxicology and chemistry. 2018;**37**(1):236–246. DOI: 10.1002/etc.3951

[114] Z Zhang H, Wang X, Qian M, Jin Y. Atrazine Exposure Induces Hepatic Metabolism Disorder in Male Adult Zebrafish. Toxics. 2022;**10**(7):400. DOI: 10.3390/toxics10070400

[115] Vaiserman A. Early-life Exposure to Endocrine Disrupting Chemicals and Later-life Health Outcomes: An Epigenetic Bridge?. Aging and Disease. 2014;**5**(6):419–29. DOI: 10.14336/AD.2014.0500419

[116] Neha V, Vijay Y, Reetesh K. A Review on Malathion Induced Toxicity in Freshwater Fishes. Flora and Fauna. 2024;**30**:127–130. DOI: 10.33451/florafauna.v30i1pp127-130

Chapter 2

Microplastics in Marine and Coastal Ecosystems: Sources, Distribution, Ecological Impact, and Mitigation Prospects

Ángeles Martínez-Orgániz, Ricardo Herrera-Navarrete and Gerson Lira-López

Abstract

Microplastic pollution is a global environmental problem of anthropogenic origin that affects marine and coastal ecosystems. Microplastics are considered emerging pollutants of great concern. Worldwide, annual plastic production amounts to 280 million tons, and a significant fraction ends up in aquatic ecosystems. Plastic fragments smaller than 5 mm are distributed in the environment through various routes, including rainwater runoff, maritime activities, and wastewater discharges. As a result of their persistence and transportability, these pollutants have toxicological effects on numerous marine species, including birds, mammals, and turtles. In this context, this chapter analyzes the main sources, distribution, ecological impact, and mitigation strategies to prevent microplastic pollution in marine environments.

Keywords: emerging pollutants, marine ecosystems, toxicological effects, human health, environmental pollution

1. Introduction

The presence of microplastics (MPs) in the marine environment has been considered an important environmental stressor [1, 2]. These pollutants have become increasingly relevant in recent years, as their presence has spread to rivers, lakes, coastal areas, oceans, and even polar regions, demonstrating their capacity for global dispersion and persistence in the environment [3]. Plastics have become the foundation of modern industry and everyday life, with annual production estimated to have grown by 9% since 1950.

The first studies on the presence of plastic waste in the oceans emerged in the 1970s and attracted greater attention from the scientific community [4–7].

The term "microplastics" was coined by Thompson et al. [8], who described them as plastic particles smaller than 5 mm, originating from plastic products and materials used daily in modern society. Between 1950 and 2021 alone, annual plastic production is estimated to have increased from 1.5 to 367 million tons [9], and in 2022, it exceeded 400 million tons. If consumption trends continue, it is expected to exceed 600 million tons by 2040 [10, 11].

Its ubiquity spans almost all industries, as it offers characteristics such as versatility, lightness, and low production costs, with applications ranging from transportation and information technology to food packaging and energy production [12–14]. Currently, more than 300 types of plastics are manufactured worldwide, of which approximately 60 are the most commonly used in commercial applications [15].

The persistence and mobility of MPs, along with their ability to be transported in marine environments, have caused toxicological effects in various marine species, including birds, mammals, and turtles. Globally, at least 267 species are affected by microplastics, including 43% of all marine mammal species, 44% of seabirds, and 86% of all sea turtle species [16]. This problem highlights their role as an emerging threat to biodiversity and the ecological integrity of aquatic ecosystems.

As a result of the impact of MP pollution on marine and coastal ecosystems, international agreements and mechanisms have been established to improve the control and management of this type of pollutant, including the Basel Convention, which, since 2019, has regulated the transboundary movement of plastic waste [17]. Similarly, the United Nations Environment Assembly (UNEA) made a historic proposal to develop a binding international legal instrument, the draft of which is expected to be completed in 2025.

While these developments reflect global recognition of the importance of establishing environmental governance in the face of this emerging problem [18], it must also be acknowledged that economic, social, and environmental challenges exist in many countries around the world. In this context, this chapter analyzes the main sources, distribution, ecological impact, and mitigation strategies to prevent microplastic pollution in marine environments.

2. Microplastics: Definition, properties, and classification

In the 1970s, the first reports on the presence of small plastic particles in marine ecosystems were published, attracting the attention of several environmental experts who sought to investigate the issue further [19]. The term "microplastics" was first introduced in a study published by Thompson et al. [8], a term that did not yet have a clear definition at that time. Another study in 2007 provided a better explanation of MP, defining them as very small plastic particles resulting from the fragmentation of larger pieces, although the size of MP was not specified [20].

The term "microplastic" was more clearly defined in 2008 at a conference organized by the US National Oceanic and Atmospheric Administration (NOAA) as tiny plastic particles, although at that time the minimum particle size was not officially specified. Currently, international standards such as ISO/TR 21960:2020, which focus on plastics and their environmental aspects, including the state of knowledge and methodologies, and ISO 4484–2:2023, which addresses materials and textile products as well as microplastics of textile origin, specifically part 2: Qualitative and quantitative analysis of microplastics, refer to particles smaller than

5 mm and larger than 1 μm as microplastics and classify those smaller than 1 μm as nanoplastics [20].

The term "plastic" refers to a range of materials created from polymers that can be molded into different shapes and sizes. It is important to note that plastics include both fossil-based and bio-based variants. Some of the main polymers are polystyrene (PS), polypropylene (PP), polyethylene (PE), polyamides (PA), polyvinyl chloride (PVC), and polyethylene terephthalate (PET) [21]. Therefore, the term "microplastic" encompasses a heterogeneous group of pollutants due to the polymers and additives used in their production [22].

Plastic products are durable, lightweight, inexpensive, and easy to handle, which is why they are widely used in our daily lives and especially in industry [23]. Microplastics are present in the environment with different physical characteristics. **Table 1** describes five different forms of microplastics.

Another way to classify plastic is by size: macroplastic (>5 mm) or microplastic (<5 mm). Microplastic can also be classified as primary if its original manufacture is <5 mm, or secondary if it is a macroplastic that has degraded due to environmental exposure [22, 25]. It is also possible to classify it based on its chemical components, of which there are 20 types. The most common MPs in the environment are polyvinyl chloride, polystyrene, polyethylene, polypropylene, and polyethylene terephthalate, among others [22].

Microplastics are also classified into two broad categories according to their origin: (1) primary and (2) secondary. Primary microplastics are classified as being less than 5 mm in size, generated by grinding or extrusion for direct use or product manufacturing. They are commonly found in medicines and household products such as beauty products, hair gel, deodorant, cosmetics, toothpaste, and water bottles. Secondary microplastics are larger plastic fragments whose biodegradation occurs due to the breakdown of bonds through the process of weathering [22, 24].

Microplastic (MP) shape	Description	Other terms used
Films	These are thin, flexible, low-density particles produced from food packaging materials and plastic bags.	Polyethylene sheet, plastic film.
Fragments	These are plastic flakes generated from the detachment of pipes or beverage bottles made of rigid synthetic polymers.	Particles, crystals, lint, dust, granules, shavings, flakes, and irregularly shaped films.
Fibers	This is a widespread plastic originating from domestic washing activities, synthetic fabrics, ropes, and fishing nets.	Filaments, microfibers, strands, and threads.
Foam	It has a porous and irregular shape, like a sponge, derived from mattresses and sheets made of synthetic rubber.	Polystyrene and expanded polystyrene.
Pellets	These are industrially produced, spherical or oval, and generally synthesized from polypropylene.	Resin pellets, granules, pre-production granules.

Source: Patidar et al. [24] and Nair and Perumal [21].

Table 1.
Categories for classifying MP by shape.

Likewise, Yadav and Mishra [2] report that the photodegradation of large plastics in marine environments is due to environmental factors such as mechanical abrasion, ultraviolet radiation, wind, and wave action, generating small plastic particles such as microplastics (<5 mm) and nanoplastics (<1,000 nm). It is important to note that academic attention has focused particularly on microplastics and nanoplastics, which are considered highly persistent pollutants with adverse effects on ecological systems and human health [26].

3. Sources and pathways of microplastics entering marine and coastal ecosystems

The accumulation of microplastics in the marine environment occurs through direct and indirect interaction with plastic waste, a critical situation that continues to expand. A common route of entry into marine ecosystems is via rivers, as they provide a pathway connecting land to sea, contaminating the ocean and the deep sea, and reaching even the most remote islands on the planet [24, 27]. Microplastics enter the marine environment through various means, such as drainage or sewer systems, and natural factors such as wind and ocean currents, among others [28].

It should be noted that microplastic pollution is concentrated in coastal areas, which are densely populated [23], despite being considered vulnerable areas due to the risks of flooding and pollution. Microplastics are shed from everyday products that are discarded daily, such as paints and facial scrubs [3]. In this scenario, it is estimated that each year more than 4.8 million tons of plastic ends up in the ocean and that 80% of the waste comes mainly from countries with high population density and limited waste management [1].

4. Distribution and environmental fate of microplastics

A 2019 report by the United Nations Industrial Development Organization (UNIDO) states that the impact of plastic pollution depends on each country's level of economic development, which is why countries with higher incomes produce more plastic waste. The economic cost of marine plastic pollution worldwide ranged from $6 billion to $19 billion in 2018 and is expected to reach $100 billion by 2040 [29]. Plastics are an integral part of everyday life due to their versatility, durability, and low cost. However, their indiscriminate use and improper disposal practices have had large-scale negative environmental impacts [26].

According to Haque and Fan [25], global plastic production was estimated at 1.3 million tons in 1950, increasing exponentially to 359 million tons in 2018, which alarms the scientific community, as by 2025 total plastic waste could reach 634 million tons, and by 2050 there will be 12 billion tons if global plastic production continues to increase under inadequate management of its final disposal [24, 25]. Globally, China is the largest producer of plastics with 59.08 tons (t) per year, while the United States contributes 37.83 t. Other countries, such as Germany, contribute 14.48 t, followed by Brazil with 11.85 t, Japan with 7.99 t, and Russia with 5.84 t [30].

Exposure to the environment can transform plastic waste into MPs. This process occurs through mechanical impact, photodegradation, environmental erosion, or microbial degradation [25], and its distribution in marine ecosystems poses a serious threat to biota [19]. However, microplastic particles have also been detected in terrestrial and freshwater environments, as well as in salt and dust [31].

Therefore, understanding the impact of MPs on various organisms is vitally important, given their widespread distribution in the environment, which makes them a problem of global concern [25]. Today, there is concern, but also growing interest in understanding the dynamics and environmental impact of microplastics in marine environments [23]. During the decade 2010–2020, there was a global increase in plastic production from 270 million tons to more than 370 million tons [2], and only 9% is recycled, 68% ends up in landfills or is incinerated, and 22% is managed improperly and ends up dispersed in the environment [32].

Microplastic pollution has raised significant concerns due to its ecological impact on marine fauna and the possibility that plastic could affect the food chain and become a serious risk to human health [9]. This situation has become a global problem due to the growing use and consumption of plastic products in activities such as agriculture, healthcare, construction, textiles, among others [22]. Microplastics are spread throughout all ocean matrices, as they constitute 95% of plastic waste [33]. In the five oceans, the concentration of MPs ranged from 0.002 to 62.50 elements/m^3, with an average abundance of 2.76 elements/m^3. The highest average concentration was found in the Atlantic Ocean (4.98 elements/m^3), while the lowest was observed in the Antarctic Ocean (0.04 elements/m^3) [34].

Since 1972, the London Convention and, later, the United Nations Convention on the Law of the Sea (UNCLOS) were the first to draw attention to the presence of plastic waste in marine ecosystems, considering it a global environmental problem due to its widespread impact on aquatic ecosystems, as they stated that most of this waste ends up in rivers, lakes, and oceans [35]. Understanding the fate of MP requires knowledge of its main transport routes, and the most common transfer routes are through soil, water, and air [25].

Nevertheless, determining the impact of MP is more complex due to the various routes it takes to reach marine ecosystems, among the most common being rainwater runoff, industrial discharges, urban wastewater, coastal sediment transport, and agricultural runoff [36]. On the other hand, air and water currents can also transport small MP particles, as thousands of tons of particles are suspended in the atmosphere, allowing them to travel to remote locations such as isolated plateaus, polar regions, and mountainous areas [37]. However, rivers remain the main route for MPs to enter the ocean [38].

Although MP pollution is concentrated in coastal areas and subtropical circulation zones, these particles have traveled to places far from the sources of pollution, such as the polar regions. This evidence has led researchers to conclude that plastic waste polluting the environment is not static but circulates from one ecosystem to another, with the smallest particles, such as MPs and nanoplastics, covering the largest areas [23, 39].

Another factor that has various effects on the fate of plastic material (PM) in the environment is climate change. This alteration in the climate can cause extreme weather events such as hurricanes, storms, and floods, which serve as transport mechanisms for the distribution of MP in the environment. Due to their natural force, these currents have the potential to transport PM over long distances, increasing the likelihood of contamination in areas far from civilization [25]. On the other hand [3], aquatic organisms can be considered a possible destination for MPs, since various studies have shown that they can be ingested by different organisms, and that in most cases they can be eliminated through excretion. However, it has also been reported that MPs remain in the tissues of the organism.

Plastic waste decomposes in soil and aquatic environments through weathering, which is an environmental process of degradation or fragmentation [40].

Due to their chemical composition, most plastics are not biodegradable in a short period of time, so prolonged time and specific environmental conditions are required for plastic material to break down into small pieces [41] of up to 5 mm, which are defined as MPs [42]. Microplastics are subject to mechanical, chemical, and biological interactions, and there is a possibility that the breakdown of MPs into smaller particles, considered to be nanoplastics, may pose greater environmental risks due to their nanometric size [38]. In this regard, high temperatures accelerate the degradation of plastic, causing small particles to be released into the environment [25].

5. Ecological and ecotoxicological impacts

The widespread presence of plastic products in modern society has led to environmental problems, as plastic waste enters the environment in various ways and accumulates mainly in marine ecosystems [23]. In this regard, scientific evidence has confirmed that marine environments are heavily polluted, and it is estimated that more than 250,000 tons of plastic waste float on the sea surface due to inadequate waste disposal methods [2, 23]. However, the estimated amount of PMs released into marine ecosystems continues to be investigated [38].

Although the most commonly found marine debris includes paper, metal, glass, and plastic, the latter is one of the most abundant and resistant to degradation due to its composition and durability [28]. Despite this, plastics break down into MPs due to the influence of various environmental factors, and these small particles pose a serious threat when distributed in aquatic ecosystems [43]. Due to their widespread dispersion and resistance to biodegradation, MP has become a serious environmental crisis [44] because of their adverse effects on the behavior, reproduction, growth, and metabolism of aquatic organisms [38].

Therefore, the impact and effect of plastic waste on marine life are highly significant due to its wide reach [28], as it travels long distances through ocean currents and wind currents [24]. Atmospheric MPs are the largest contributors to the oceans after continental rivers [45]. Plastic waste is present in marine environments from the sea surface to the seabed [27] and has also impacted marine organisms [28], such as plankton, zooplankton, coral reefs, fish, mammals, and seabirds [23, 32]. Despite the attention given to scientific research, plastic pollution has been on the rise over the last decade [32], with research focusing on marine species and only about 4% addressing studies in freshwater systems [44].

Furthermore, numerous aquatic organisms have been reported to ingest PM directly, including sea urchins, mussels, crabs, barnacles, copepods, sandworms, and amphipods [2]. Some coastal regions have also shown MP contamination in commercial fish such as tilapia (*Oreochromis niloticus*), moontail bullseye (*Priacanthus hamrur*), longnose trevally (*Carangoides chrysophrys*), long-whiskered catfish (*Sciades sona*), pama croaker (*Otolithoides pama*), Bombay duck (*Harpadon nehereus*), neglected grenadier anchovy (*Coilia neglecta*), hairfin anchovy (*Setipinna tenuifilis*), short-bodied sardine (*Sardinella brachysoma*), among others [46].

It is important to highlight that the ingestion of MP by marine organisms is accidental, as they mistake them for food. These particles are found in sediments or water columns and tend to translocate to the intestines of many organisms, where they can remain for relatively long periods and subsequently reach humans through the consumption of foods such as fish or shellfish [37]. Therefore, the existence of MP in marine ecosystems is evident and represents a latent threat. MP contamination has even been reported in more than 160 marine species and 39 freshwater species [22].

The ingestion of MP can affect the digestive system of aquatic organisms, as physical obstruction causes a reduction in food intake, negatively affecting growth and reproduction. This physiological impact can cause a decline in the species' population, mainly in species threatened by environmental changes [44] or poachers. However, the ingestion of MP in the species can block or damage the digestive tract, as well as cause physical and chemical toxicity [37].

Disruption of the digestive system or other vital biological processes due to MP intake poses a threat to all biota [20]. Microplastics accumulate in the vital organs of fish (gills, intestines, and liver) and influence their predatory performance, as they cause confusion when obtaining food. The accumulation of MP causes obstruction, behavioral changes, histopathological changes in the intestines, and liver metastasis in fish, as well as growth inhibition, endocrine disruption, and stress. Likewise, various studies have shown a range of effects, from minor alterations in physiological functions to serious adverse effects that can cause death in organisms contaminated by MP (**Table 2**) [19].

Microplastic pollution is an emerging issue of global concern due to its widespread presence in the environment, as it represents a source of chemicals that are hazardous to aquatic organisms due to its composition and sorption properties [38]. It should be noted that UV radiation, heat, and chemical reactions are considered weathering mechanisms, which cause the decomposition, fragmentation, or degradation of plastics, breaking down long-chain polymers into smaller polymers [25].

Species name	Effects
Blue mussel (*Mytilus edulis*)	Destabilization of lysosomal membranes and formation of granulomas in the digestive glands.
Mytilus galloprovincialis	Ingestion of resin pellets can cause intestinal obstruction, reduced feeding, inflammation, and other serious effects.
Zooplankton	Accumulation of phthalic acid esters through ingestion affects the reproductive, hepatic, renal, and nervous systems.
Blue mussels (*Mytilus galloprovincialis*)	The absorption of pyrene ingested by blue mussels causes neurotoxic effects and lethal effects on the organism's DNA.
Pelagic fishes and holothurians	Ingestion of plastic pellets by sea cucumbers can reach other organisms through the food chain.
Copepod (*Calanus helgolandicus, C. cristatus, Euphausia pacifica*)	Decreased reproduction rates, reduced feeding, and decreased egg production.
European flat oysters (*Ostrea edulis*)	Ingestion and abnormal respiration rates.

Source: Onyena et al. [47].

Table 2.
Effects of MP on marine organisms.

The results show a direct correlation between metals and MPs found in some species, mainly fish, indicating that the metals were absorbed by this type of plastic pollutant [20]. In this sense, in addition to their resistance to biodegradation and wide dispersion, MPs can adsorb and carry heavy metals, thus intensifying their toxic effects [44], acting as a transfer route for organic pollutants and heavy metals through aquatic species [19], as well as causing oxidative stress, cell damage, and other effects [22]. Another factor that accelerates the degradation of plastics is global warming. At higher temperatures, plastics decompose at an accelerated rate, thus releasing very small microplastic particles [25].

Studies have shown that smaller MPs and nanoplastics, particularly nanoplastics, are more toxic than larger MPs [35]. Microplastics usually take several months to degrade. This variation is due to the structure of the polymer. However, the degradation process can be further prolonged if the particle is not affected by other environmental factors of a mechanical, biological, or chemical nature. For example, while plastic decomposes quickly in marshes, it has been observed to take longer in sludge and soil [20].

Plastics are characterized by their strength, low cost, lightness, and chemical inertia, which have led to an increase in their production. However, these properties hinder the degradation process [2]. Microplastics are not only generated from the degradation process of larger plastic waste but can also be intentionally manufactured in small sizes for various industrial applications. Due to their tiny size, aquatic organisms can easily ingest them, which can lead to bioaccumulation and possible trophic transfer within networks [26].

Although MPs are potentially consumable by a variety of organisms, mainly commercial fish, they can accumulate in the body of the organism and affect various trophic levels, as they could move widely throughout the food web [19]. Most everyday products are made of plastic. During their manufacture, various solvents and other chemicals are used as catalysts, in addition to various additives such as stabilizers, pigments, and retardants. Plastic must have specific characteristics such as strength, color, and flexibility; however, when it degrades, it can release chemical additives or leach into the environment with far-reaching repercussions. Therefore, MPs pose a risk to food safety, mainly in organisms at lower trophic levels [21].

Recent research confirms that MPs pose a serious threat to organisms inhabiting terrestrial and marine ecosystems, as they compromise their biodiversity and function [38]. In addition, they can negatively affect the food web in marine environments [43]. It has been shown that MP pollution can cause physical changes in habitats, with coral reefs among the most affected [28]. Most research on the environmental impacts of MPs has been conducted in marine environments, while studies in freshwater ecosystems have been very limited [44].

Even so, various disciplines have studied the impact of PM pollution, such as biology, chemistry, environmental toxicology, and microbial ecology, among others [9]. In this context, several case studies have been documented in recent years. A study of the Plankenburg River in the Western Cape of South Africa investigated the presence of MP in water and sediment samples. The study provided an understanding of the occurrence of microplastics in the river system and established a baseline for comparing future research with the assessment of water and sediments in South Africa's freshwater systems [38].

It is important to note that rivers usually flow into the ocean, which also implies marine pollution by MPs. Another study examined the pathways of MP and nanoplastic pollution, as well as their interactions with other environmental stressors that affect marine food webs, such as chemical pollution and climate change [25]. This highlights an urgent need to promote research on the behavior and fate of MPs and nanoplastics [35].

In this same approach, a study was conducted on a segment of the Rupsha River in Bangladesh, which aimed to assess the extent of microplastic pollution and its impacts on fish species in relation to human consumption [46]. Other studies were highlighted by Haque and Fan [25], which linked plastic pollution to climate change, eutrophication, anthropogenic factors, and, above all, human health. However, their main contribution focused on discussing the fate of microplastics in all types of environments (atmosphere, hydrosphere, and pedosphere) and under the influence of climate change [25].

Gliaudelytė et al. [20] revealed a study conducted in a marine protected area in the Ross Sea, finding textile- and synthetic-based MPs in marine snails (*Neobuccinum eatoni*). The investigation into the origin of the MP pointed to the research station, where it was verified that the type of MP matched the material of the scientists' technical clothing, leading to the conclusion that it was the wastewater used to wash the technical clothing. This research led to other, more specific studies in which samples were collected from six different points in coastal regions to compare them with the microplastics found in filters from domestic washing machines in the area of influence, concluding that they had a similar composition.

6. Human exposure and potential health risks of MP

Over the last decades, MPs have become one of the biggest environmental problems and challenges in current research. Their global impact threatens not only the environment but also human health, due to their transfer through food and human exposure to various plastic products [24, 48]. In fact, researchers have found plastic particles in rainwater, drinking water, and even in the human placenta, lungs, and blood [20, 32]. Consumption of MP particles puts humans at risk of cytotoxicity, hypersensitivity, hemolysis, neurological conditions, and other diseases such as cancer [24].

Microplastics are so small that traces have also been found in a wide variety of foods and beverages, such as salt, sugar, and beer, which has alarmed the scientific community, as MPs enter the human body directly through ingestion, in addition to inhalation or skin contact [20]. It is also important to note that the human body can eliminate MPs through the biliary and digestive systems. However, MPs are transported through the bloodstream and accumulate in 15 different biological components of the human body, such as the lungs, liver, spleen, colon, and feces, including the placenta and breast milk [1].

In this regard, the inadequate management of certain human activities could introduce various xenobiotic pollutants into aquatic ecosystems, either directly or indirectly [19], such as swimming, boating, tourism, fishing, and industrial operations (oil production, aquaculture, etc.) [24]. Food and beverage containers, fishing gear, and plastic bags are the most common types of waste found in marine waters, which travel through sewage systems, reach rivers, and eventually wash up on beaches [19]. The most polluted areas are coastal regions due to human activity aimed at boosting the economy. However, tourism and migration are the most significant factors in this environmental dynamic [24].

Studies show that MPs are a big threat to our health. They can get into our bodies through touch, breathing, or the food chain. However, MPs can cause cellular toxicity and lead to kidney, bladder, prostate, or other chronic diseases, as well as nerve damage and brain dysfunction [49].

7. Current mitigation strategies and future prospects

Solving the problems posed by MP requires knowledge from various disciplines to obtain more accurate data on the distribution, behavior, and impacts of MP, as it is considered a global threat that impacts the pedosphere, hydrosphere, and atmosphere [25]. Considering this situation, some countries around the world are attempting to regulate the production, distribution, and consumption of plastics. For example, the European Commission is implementing initiatives such as the Circular Economy Action Plan, while others are applying more rigorous strategies, such as Directives on packaging and packaging waste and the Waste Framework Directive, which seek to ban single-use plastics [32].

In addition, the species included in the Red List of the International Union for Conservation of Nature (IUCN) is another strategy that justifies special attention to marine biodiversity with regard to MP pollution, as the problem has been recognized in global agreements carried out at the 10th, 11th, and 12th Conferences of the Parties to the Convention on the Conservation of Migratory Species of Wild Animals (CMS Resolutions 10.4, 11.4, and 13.4), as well as in the 11th Conference of the Parties to the Convention on Biological Diversity (CBD COP 11 Decision XI/18), and the European Union's Marine Strategy Framework Directive (MSFD) [28].

Plastic pollution can also be reduced by adopting various R strategies. However, innovative strategies are essential, but they require a comprehensive understanding of the main causes, as well as the role of global sustainability agendas in addressing this serious problem. These strategies are based on methods grounded in the principles of the circular economy, the SDGs, and international standards such as ISO 59000, which addresses the circular economy framework [32].

Other strategies to reduce the presence of MP in marine ecosystems and protect the coastal environment include promoting and implementing the 3 R policy (Reduce, Reuse, and Recycle), as well as public awareness initiatives through educational campaigns on the effects of plastic pollution, to promote a collective understanding of the far-reaching consequences and encourage sustainable practices [30].

Furthermore, solving MP problems requires diverse knowledge and approaches.

For example, the planetary boundaries approach to mitigating plastic pollution already assesses at least six of the nine fundamental processes addressed by this scientific framework: climate change, biogeochemical flows, freshwater change, loss of biosphere integrity, terrestrial system change, and novel entities, all of which are related to plastic pollution [50]. While the use of this approach is very promising, obtaining accurate data on the behavior, fate, and impacts of MPs is complex due to the diverse characteristics and composition of plastic waste [25].

Another safer option for eliminating MPs, due to their unique metabolic mechanism, is the microbial approach. There are more than 400 microorganisms capable of degrading MPs, among which fungi stand out as the best-performing plastic degraders among microbes, due to the release of unique enzymes as well as their ability to thrive on hazardous components. However, there are still gaps in our knowledge about the mechanisms underlying biodegradation [2].

In general, various strategies have had favorable results in waste management and continue to lead to the mitigation of plastic pollution. One example of this is the European policy promoted by the "Strategy for Plastics in a Circular Economy," which aims to establish a new culture that promotes and encourages the reuse and recycling of plastic materials. Although various strategies have been implemented to clean up rivers, lakes, and oceans, efforts to combat the enormous extent of microplastic pollution have so far proved insufficient [32].

Policies, legislation, and plastic waste management play a key role when applied strategically against environmental pollution, particularly that caused by MPs [32]. There are also plastic restriction policies that can help mitigate the impacts. Countries should focus their environmental policies on the production process rather than the consumption phase, as this can be a key to exercising better control. However, the Malaysian government has taken a different approach, increasing public participation in controlling plastic pollution through incentive measures, while Pakistan has effectively reduced plastic bags by more than three-quarters through policy interventions, and Rwanda has implemented a plastic ban driven by the tourism economy [49].

According to Bhuyan et al. [30], with the prohibition of single-use plastics and the reduction of pollution sources, it will be very difficult for MPs to reach the environment. It is also possible to achieve a significant reduction in plastic products in industry by designing eco-friendly products and adopting alternatives. In addition, regulations can lead to a more circular and efficient approach to resource use through research, innovation, and the adoption of sustainable practices to reduce MP [32]. The prohibition of single-use plastics is also a strategy that has gained relevance in various countries, especially when they commit to integrating it into their public policies in order to mitigate MP pollution [47].

Biodegradable MPs are emerging as a potential alternative to conventional plastics. Despite being a promising solution, biodegradable MPs contribute to water pollution, creating complex problems, as studies have found MPs of this type in drinking water supplies [1]. Microplastics are considered emerging pollutants of great concern, and there is little or no awareness in the public and private sectors about the threat these small particles pose to the environment. Therefore, reducing the entry of microplastics into the aquatic environment involves taking into account all possible sources [3], as well as collaboration and partnerships with society, government, and academia.

8. Conclusions

Microplastic pollution in marine ecosystems is increasing, and interest in understanding and analyzing the main sources of MP, its distribution, and ecological impact has led to the search for mitigation strategies to prevent pollution from this emerging contaminant. It, therefore, remains and will continue to be one of the major environmental problems to be solved in the coming decades, requiring a global effort on the same scale as global warming or biodiversity loss.

Although most research focuses on marine environments and there is very little evidence in freshwater ecosystems, much remains to be learned about their ecological and ecotoxicological impacts, considering not only their increasing accumulation but also the diversity of chemical compounds that constitute them, in addition to associated toxic compounds (e.g., heavy metals). In addition, PM present in marine and coastal ecosystems also poses a significant risk to human health.

To date, regulatory strategies implemented by some countries (mainly in the European Union) have significantly reduced the accumulation of MPs in marine and coastal ecosystems. On the other hand, the promotion and implementation of the 3 R policy (Reduce, Reuse, and Recycle) do not seem to be sufficient in the context of current population growth and the demand for plastics for daily activities.

The adoption of innovative R strategies, beyond the 3 R model, based on circular economy principles, accompanied by public awareness initiatives through educational campaigns and the promotion of sustainable practices, is an essential strategy, allowing time for scientific research to design plastics with much greater biodegradability or to develop organisms that reduce the accumulation of plastics in aquatic environments.

Therefore, institutional and academic partnerships, and, above all, citizen participation, are important strategies for mitigating PM pollution. Combined with current technology, they can achieve progress and respond favorably to the global commitment of the 2030 Agenda.

Author details

Ángeles Martínez-Orgániz[1], Ricardo Herrera-Navarrete[2*] and Gerson Lira-López[3]

1 Autonomous University of Guerrero, School of Natural Sciences, Chilpancingo, Mexico

2 Autonomous University of Guerrero, Regional Development Science Center, Acapulco, Mexico

3 Autonomous University of Guerrero, Faculty of Biological Chemical Sciences, Chilpancingo, Mexico

*Address all correspondence to: rherrera@uagro.mx

References

[1] Bexeitova K, Baimenov A, Varol EA, Kudaibergenov K, Zhantikeyev U, Sailaukhanuly Y, Toshtay K, Tauanov Z, Azat S, Berndtsson R. Microplastics in freshwater systems: A review of classification, sources, and environmental impacts. Chemical Engineering Journal Advances. 2024;**20**:100649. DOI: 10.1016/j.ceja.2024.100649

[2] Yadav P, Mishra V. Comprehending microplastic pollution in diverse environment: Assessing fate, impacts, and remediation approaches. International Biodeterioration & Biodegradation. 2025;**196**:105953. DOI: 10.1016/j.ibiod.2024.105953

[3] Auta HS, Emenike CU, Fauziah SH. Distribution and importance of microplastics in the marine environment: A review of the sources, fate, effects, and potential solutions. Environment International. 2017;**102**:165–176. DOI: 10.1016/j.envint.2017.02.013

[4] Carpenter EJ, Smith Jr KL. Plastics on the Sargasso sea surface. Science. 1972;**175**:1240–1241. DOI: 10.1126/science.175.4027.1240

[5] Colton JB, Burns BR, Knapp FD. Plastic particles in surface waters of the Northwestern Atlantic. Science. 1974;**185**:491–497. DOI: 10.1126/science.185.4150.491

[6] Fowler CW. Marine debris and northern fur seals: A case study. Marine Pollution Bulletin. 1987;**18**:326–335. DOI: 10.1016/S0025-326X(87)80020-6

[7] Coe JM, Rogers DB. Marine Debris: Sources, Impacts, and Solutions. Nueva York, NY: Springer; 1997:5–6. DOI: 10.1007/978-1-4613-8486-1_1

[8] Thompson RC, Olsen Y, Mitchell RP, Davis A, Rowland SJ, John AWG, McGonigle D, Russell AE. Lost at Sea: Where is all the Plastic? Science. 2004;**304**:838–838. DOI: 10.1126/science.1094559

[9] Bakhshaee A, Babakhani P, Ashiq MM, Bell K, Salehi M, Jazaei F. Potential impacts of microplastic pollution on soil–water–plant dynamics. Scientific Reports. 2025;**15**:9784. DOI: 10.1038/s41598-025-93668-0

[10] Geyer R, Jambeck JR, Law KL. Production, use, and fate of all plastics ever made. Science Advances. 2017;**3**:e1700782. DOI: 10.1126/sciadv.1700782

[11] OECD. Global Plastics Outlook: Economic Drivers, Environmental Impacts and Policy Options. Paris: OECD Publishing; 2022

[12] Abedsoltan H. Applications of plastics in the automotive industry: Current trends and future perspectives. Polymer Engineering & Science. 2023. DOI: 10.1002/pen.26604

[13] Agarwal S, Gupta RK. 29 - Plastics in Buildings and Construction. In: Kutz M, editor. Applied Plastics Engineering Handbook. 2nd ed. William Andrew Publishing; 2017. DOI: 10.1016/B978-0-323-39040-8.00030-4pp. 635-649

[14] Jia C, Das P, Kim I, Yoon Y-J, Tay CY, Lee J-M. Applications, treatments, and reuse of plastics from electrical and electronic equipment.

Journal of Industrial and Engineering Chemistry. 2022;**110**:84–99. DOI: 10.1016/j.jiec.2022.03.026

[15] Zhang Q, Xu EG, Li J, Chen Q, Ma L, Zeng EY, Shi H. A review of microplastics in table salt, drinking water, and air: Direct human exposure. Environmental Science & Technology. 2020;**54**:3740–3751. DOI: 10.1021/acs.est.9b04535

[16] Samiul Islam FA. The effects of plastic and microplastic waste on the marine environment and the ocean. European Journal of Environment and Earth Sciences. 2025;**6**:1–9. DOI: 10.24018/ejgeo.2025.6.3.508

[17] Rummel-Bulska I. Chapter II.2 - The Basel Convention and its implementation. In: Twardowska I, editor. Waste Management Series. Vol. **4**. Elsevier; 2004. p. 133–169. DOI: 10.1016/S0713-2743(04)80009-6

[18] UNEP. From Pollution to Solution: A global assessment of marine litter and plastic pollution. Availabe from: https://www.unep.org/resources/pollution-solution [Accessed: 2025-September-02]

[19] Bagheri T, Gholizadeh M, Abarghouei S, Zakeri M, Hedayati A, Rabaniha M, Aghaeimoghadam A, Hafezieh M. Microplastics distribution, abundance and composition in sediment, fishes and benthic organisms of the Gorgan Bay, Caspian Sea. Chemosphere. 2020;**257**:127201. DOI: 10.1016/j.chemosphere.2020.127201

[20] Gliaudelytė U, Persson M, Daukantienė V. Impact of textile composition, structure, and treatment on microplastic release during washing: A review. Textile Research Journal. 2024;**95**:220–232. DOI: 10.1177/00405175241260066

[21] Nair HT, Perumal S. Trophic transfer and accumulation of microplastics in freshwater ecosystem: Risk to food security and human health. International Journal of Ecology. 2022;**10**:1234078. DOI: 10.1155/2022/1234078

[22] Singh S, S S, Sachdev S, Sahoo KS, Ambade B, Bauddh K. Microplastic pollution in agricultural environments: Origins, impacts, and mitigation strategies. Physics and Chemistry of the Earth, Parts A/B/C. 2025;**138**:103866. DOI: 10.1016/j.pce.2025.103866

[23] Zhang W, Zhang S, Wang J, Wang Y, Mu J, Wang P, Lin X, Ma D. Microplastic pollution in the surface waters of the Bohai Sea, China. Environmental Pollution. 2017;**231**:541–548. DOI: 10.1016/j.envpol.2017.08.058

[24] Patidar K, Alshehri M, Singha W, Alrasheedi M, Younis AM, Dumka UC, Ambade B. Assessing the microplastic pandemic: Prevalence, detection, and human health impacts in Asian aquatic environments. Physics and Chemistry of the Earth, Parts A/B/C. 2025;**137**:103800. DOI: 10.1016/j.pce.2024.103800

[25] Haque F, Fan C. Fate of microplastics under the influence of climate change. iScience. 2023;**26**:107649. DOI: 10.1016/j.isci.2023.107649

[26] Sebteoui K, Csabai Z, Stanković J, Baranov V, Jovanović B, Milošević D. Downsizing plastics, upsizing impact: How microplastic particle size affects Chironomus riparius bioturbation activity. Environmental Research.

2025;**270**:121055. DOI: 10.1016/j.envres.2025.121055

[27] Van Cauwenberghe L, Vanreusel A, Mees J, Janssen CR. Microplastic pollution in deep-sea sediments. Environmental Pollution. 2013;**182**:495–499. DOI: 10.1016/j.envpol.2013.08.013

[28] Gall SC, Thompson RC. The impact of debris on marine life. Marine Pollution Bulletin. 2015;**92**:170–179. DOI: 10.1016/j.marpolbul.2014.12.041

[29] UNEP. Marine plastic debris and microplastics – global lessons and research to inspire action and guide policy change. Availabe from: https://www.unep.org/resources/publication/marine-plastic-debris-and-microplastics-global-lessons-and-research-inspire [Accessed: 2025-September-03]

[30] Bhuyan MS, Senapathi V, Karthikeyan S, Kunda M, Islam MT, Iheanacho S, Choudhury AK. Microplastics contamination in seaweed: Impacts on human health and mitigation approaches. Discover Oceans. 2025;**2**:2. DOI: 10.1007/s44289-025-00041-0

[31] Veerasingam S, Ranjani M, Venkatachalapathy R, Bagaev A, Mukhanov V, Litvinyuk D, Verzhevskaia L, Guganathan L, Vethamony P. Microplastics in different environmental compartments in India: Analytical methods, distribution, associated contaminants and research needs. TrAC Trends in Analytical Chemistry. 2020;**133**:116071. DOI: 10.1016/j.trac.2020.116071

[32] Kyriakopoulos GL, Zorpas AA, Inglezakis VJ, Rodríguez-Barroso R. The silent invaders: Microplastic accumulation, impacts, and monitoring approaches. Toxics. 2025;**13**:192. DOI: 10.3390/toxics13030192

[33] Zhang S, Wang W, Yan P, Wang J, Yan S, Liu X, Aurangzeib M. Microplastic migration and distribution in the terrestrial and aquatic environments: A threat to biotic safety. Journal of Environmental Management. 2023;**333**:117412. DOI: 10.1016/j.jenvman.2023.117412

[34] Mutuku J, Yanotti M, Tocock M, Hatton Macdonald D. The abundance of microplastics in the world's oceans: A systematic review. Oceans. 2024;**5**:398–428. DOI: 10.3390/oceans5030024

[35] Nawaz F, Islam ZU, Ghori SA, Bahadur A, Ullah H, Ahmad M, Khan GU. Microplastic and nanoplastic pollution: Assessing translocation, impact, and mitigation strategies in marine ecosystems. Water Environment Research. 2025;**100**:70032. DOI: 10.1002/wer.70032

[36] Li Y, Lu Z, Zheng H, Wang J, Chen C. Microplastics in surface water and sediments of Chongming Island in the Yangtze Estuary, China. Environmental Sciences Europe. 2020;**32**:15. DOI: 10.1186/s12302-020-0297-7

[37] Abbasi S, Soltani N, Keshavarzi B, Moore F, Turner A, Hassanaghaei M. Microplastics in different tissues of fish and prawn from the Musa Estuary, Persian Gulf. Chemosphere. 2018;**205**:80–87. DOI: 10.1016/j.chemosphere.2018.04.076

[38] Apetogbor K, Pereao O, Sparks C, Opeolu B. Spatio-temporal distribution of microplastics in water and sediment samples of the Plankenburg river,

Western Cape, South Africa. Environmental Pollution. 2023;**323**:121303. DOI: 10.1016/j.envpol.2023.121303

[39] Lwanga EH, Beriot N, Corradini F, Silva V, Yang X, Baartman J, Rezaei M, van Schaik L, Riksen M, Geissen V. Review of microplastic sources, transport pathways and correlations with other soil stressors: A journey from agricultural sites into the environment. Chemical and Biological Technologies in Agriculture. 2022;**9**:20. DOI: 10.1186/s40538-021-00278-9

[40] Arp HPH, Kühnel D, Rummel C, MacLeod M, Potthoff A, Reichelt S, Rojo-Nieto E, Schmitt-Jansen M, Sonnenberg J, Toorman E, et al. Weathering plastics as a planetary boundary threat: Exposure, fate, and hazards. Environmental Science & Technology. 2021;**55**:7246–7255

[41] Yang CZ, Yaniger SI, Jordan VC, Klein DJ, Bittner GD. Most plastic products release estrogenic chemicals: A potential health problem that can be solved. Environmental Health Perspectives. 2011;**119**:989–996. DOI: 10.1289/ehp.1003220

[42] Jiang N, Luo W, Zhao P, Ga B, Jia J, Giesy JP. Distribution of microplastics in benthic sediments of Qinghai Lake on the Tibetan Plateau, China. The Science of the Total Environment. 2022;**835**:155434. DOI: 10.1016/j.scitotenv.2022.155434

[43] Rasta M, Khodadoust A, S. Taleshi M, S. Lashkaryan N, Shi X. Potential use of gammarus (Pontogammarus maeoticus) and shrimp (Palaemon elegans) as biomonitors of microplastics pollution in coastal environments. Environmental Pollution. 2024;**362**:124959. DOI: 10.1016/j.envpol.2024.124959

[44] Najibzadeh M, Kazemi A, Esmaeilbeigi M, Hassan HU. Impact of microplastic contamination derived from human activities on anuran species. Process Safety and Environmental Protection. 2025;**197**:107002. DOI: 10.1016/j.psep.2025.107002

[45] Xu L, Tian X, Bai X, Li K, Zhang G, Zhang M, Rillig MC, Huang Y, Hu M. Atmospheric microplastic input into wetlands: Spatiotemporal patterns, drivers, and unique ecological impacts. Water Research. 2025;**268**:122601. DOI: 10.1016/j.watres.2024.122601

[46] Shakik A, Brohomo P, Kabir MS, Islam MS, Mizan MH. Microplastic contamination in Rupsha River of Bangladesh and its impacts on fish species. Regional Studies in Marine Science. 2025;**85**:104130. DOI: 10.1016/j.rsma.2025.104130

[47] Onyena AP, Aniche DC, Ogbolu BO, Rakib MR, Uddin J, Walker TR. Governance strategies for mitigating microplastic pollution in the marine environment: A review. Microplastics. 2022;**1**:15–46. DOI: 10.3390/microplastics1010003

[48] Puhar J, Sturm MT, Myers E, Schober D, Korzin A, Vujanović A, Schuhen K. When technology meets sustainability: Microplastic removal from industrial wastewater, including impact analysis and life cycle assessment. Water. 2025;**17**:671. DOI: 10.3390/w17050671

[49] Zhang L, Wang Y, Xu Z, Chen L. Quantitative Study on the Synergistic Effect of China's Plastic Restriction Policy from 2008 to 2025. Sustainability. 2025;**17**. DOI: 10.3390/su17167355

[50] Villarrubia-Gómez P, Carney Almroth B, Eriksen M, Ryberg M, Cornell SE. Plastics pollution exacerbates the impacts of all planetary boundaries. One Earth. 2024;7:2119–2138. DOI: 10.1016/j.oneear.2024.10.017

Chapter 3

Pharmaceuticals as Emerging Contaminants in the Environment: Impacts and Remediation Processes

Harry Lik Hock Lau, Rusydi R. Sofian, Syahirah Nabilah Aedy Aewandy, Nur Diana Bazilah Awang Idris, Nur Aisyah Abdul Munir, Nur Nabaahah Roslan, Aisyah Farhanah Abdul Majid, Anwar Usman and Muhammad Nur

Abstract

Pharmaceutical active compounds (PhACs) are an increasingly prevalent and growing class of emerging contaminants that are continuously accumulating in waste water in proportion to global population growth. Most PhACs enter the environment through domestic, agricultural, healthcare, and industrial sources, primarily via waste water treatment plants, which act as the main pathways to surface water bodies. PhACs in waste water can undergo complex transformations through abiotic and biotic processes, resulting in numerous derivative compounds. These PhACs are potentially toxic to aquatic organisms, plants, and ecosystems, and contribute to the development of antimicrobial resistance, which poses a significant threat to human health. This chapter provides an overview of the bio-toxicity, health risks, detection methods, and effective removal methods for PhACs in waste water and water systems. It highlights the challenges of detecting trace levels of PhACs using highly sensitive techniques, such as chromatography, spectroscopy, mass spectrometry, electrochemical and optical methods, immunoassays, and microbial biosensors. Furthermore, it discusses promising remediation techniques – such as membrane filtration, ion exchange, adsorption, coagulation–flocculation, reduction, electrochemical methods, advanced oxidation processes, activated sludge, and constructed wetlands – along with their respective advantages and limitations. Finally, the chapter presents future perspectives on developing integrated or hybrid remediation approaches aimed at improving removal and degradation efficiency, cost-effectiveness, and overall economic feasibility. It also emphasizes the need for comprehensive evaluations of the toxicity and risks associated with by-products formed during the treatment of PhACs.

Keywords: pharmaceutical waste, emerging contaminants, waste water, waste water treatment, remediation

1. Introduction

Pharmaceuticals are compounds synthesized and used to alleviate symptoms of illnesses, prevent and diagnose chronic diseases, and restore the functions of organs. Pharmaceutical active compounds (PhACs), the biologically active components of pharmaceutical products, are designed to be stable and to interact efficiently with specific target molecules in the body. However, a substantial portion of PhACs is not metabolized in humans or animals, leading to their accumulation in sewage treatment plants and, ultimately, in water resources. In the environment, PhACs can undergo complex transformations through abiotic and biotic processes, forming numerous derivative compounds that may negatively impact aquatic organisms and ecosystems [1]. For instance, recent studies have reported carcinogenic and mutagenic effects of prolonged exposure to various PhACs in aquatic organisms, such as *Danio rerio* (zebrafish), *Daphnia magna* (water flea), *Aliivibrio fischeri* (bioluminescent bacteria), and *Lemna minor* (duckweed), highlighting the serious threat these compounds pose to human health, particularly through trophic transfer [2, 3]. In humans, several adverse health effects associated with exposure to PhACs include liver damage, kidney cancer, testicular cancer, thyroid disease, and developmental issues in fetuses [4].

The increasing consumption and global market growth of pharmaceuticals, projected at a rate of 5.8% per year, along with global population growth, have been attributed to rising levels of a wide range of PhACs and their metabolites in the environment [5]. In this context, assessments of the presence, dispersion mechanisms, and environmental impacts of PhACs have been conducted in various countries worldwide. According to recent reports, 25% of water samples from more than 200 rivers contained at least one PhAC, with the highest concentrations detected in rivers located in countries with poor waste water treatment facilities. In another study, more than 700 PhACs were detected in surface waters, with concentrations typically in the range of a few tens to a few thousand micrograms per liter [6]. Although these concentrations are much lower than those of other emerging contaminants, and the toxicological properties of individual PhACs at low concentrations may not be harmful, their mixtures can have serious impacts, causing adverse effects on aquatic organisms.

The presence of pharmaceuticals as emerging contaminants in waste water is not only due to the ingestion and excretion of PhACs and their metabolites, but also to unmonitored over-the-counter medications and anthropogenic activities such as improper disposal of expired pharmaceuticals, illegal sewage discharge, and inefficient waste water treatment processes. The complex composition of waste water, containing various organic wastes with different salinity and pH levels, makes the degradation or removal of PhACs using conventional waste water treatment methods such as chlorination, filtration, flocculation, or sedimentation even more difficult and labor-intensive, as these conventional methods are primarily designed to remove organic matter and pathogens, not PhACs [7]. As a result, the persistence of PhACs in aquatic systems appears inevitable and has become a growing environmental concern.

Considering the environmental concerns, the degradation or removal of PhACs from waste water requires multidimensional strategies and remains a significant challenge for the scientific community, environmental agencies, and regulatory authorities. The primary goals are to reduce the release of pharmaceuticals into water systems, prevent the improper disposal of expired medications, and remove existing PhACs from the environment [8]. Therefore, all innovative strategies should be based on a comprehensive understanding of the environmental and human health impacts of PhACs, the various sources and transport pathways of these emerging contaminants, the continuous monitoring of aquatic systems, and the development of effective remediation methods. In this context, this chapter is organized as follows: Section 2 outlines the sources and environmental pathways of PhACs; Section 3 discusses their negative impacts on the environment; Section 4 summarizes the methods used to detect and identify trace amounts of PhACs; Section 5 presents a detailed overview of waste water remediation techniques; and Section 6 proposes future perspectives on designing integrated or hybrid remediation approaches. These insights aim to deepen the understanding of the environmental impacts and treatment strategies for the removal of PhACs.

2. Sources and environmental pathways

The primary sources of PhACs in the environment are pharmaceuticals used for human and animal health. These compounds enter the environment through several routes, including domestic households, healthcare facilities, pharmaceutical industries, and agricultural and veterinary operations such as livestock farms, poultry farms, and aquaculture systems. **Figure 1** summarizes the sources, environmental pathways, and entry routes of PhACs and their metabolites into water resources.

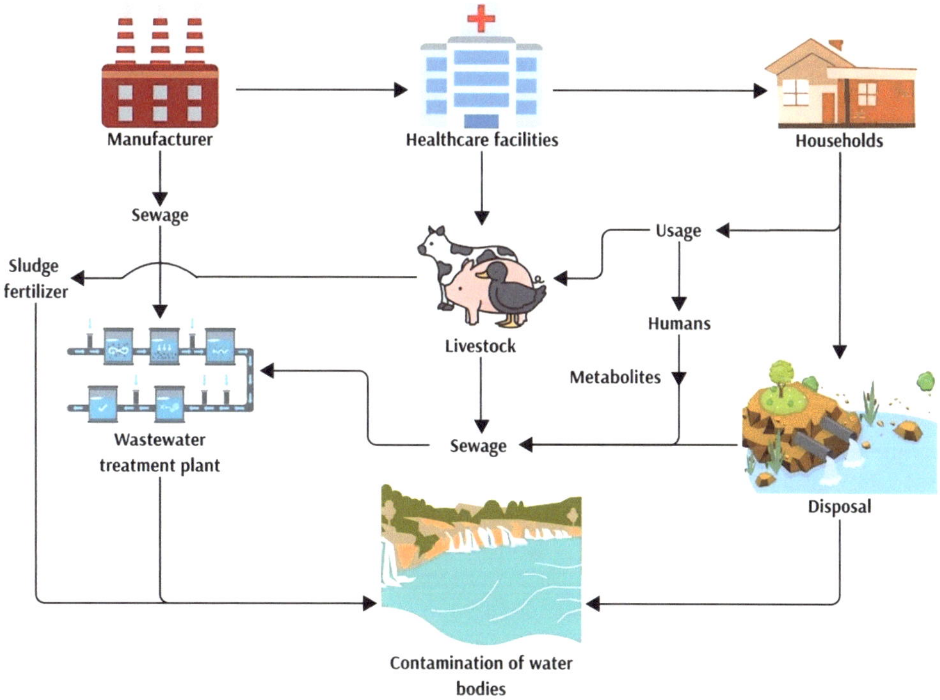

Figure 1.
Sources and pathways of PhACs in the environment, including metabolites, sludge fertilizers, improper disposal, waste water treatment plants, and finally the flow of the emerging contaminants into water systems.

PhACs are present in liquid waste, liquid effluent, and solid waste. Domestic households contribute to PhAC contamination through the discharge of liquid and solid waste into sewage systems, as well as the improper disposal of expired or unused medications. Several studies have shown that, on average, 74% of 60 different PhACs are excreted after consumption. Furthermore, surveys indicate that 89% of the public in the United States and 68% in the United Kingdom improperly discard pharmaceuticals via drains, sinks, toilets, or general non-hazardous waste [9].

Healthcare facilities, such as hospitals, generate PhACs through medical waste resulting from healthcare activities. Pharmaceutical manufacturing plants release waste and effluents containing undesired residues of PhACs with high biological activity and chemical reactivity into the environment. Similarly, livestock, poultry, and aquaculture operations frequently administer pharmaceuticals and veterinary medicines to animals. Since these animals do not fully metabolize the medications, a substantial portion of PhACs is excreted in manure. Improper treatment and discharge of this manure subsequently introduce PhACs into the environment. As a result, it is common for PhACs to end up in surface water systems [10].

The main environmental entry routes of PhACs into water resources include waste water treatment plants (WWTPs), runoff from landfill leachate, and the application of sewage sludge in agriculture. WWTPs serve as the final checkpoint where waste water from domestic households, healthcare facilities, and pharmaceutical industries converges. Therefore, various physical, chemical, or biological treatment methods are employed to separate and remove different types of contaminants, with varying degrees of effectiveness. Using conventional waste water treatment methods, PhACs and other organic compounds can be degraded depending on their polarity and water solubility. However, many of these emerging contaminants escape from WWTPs into the environment. Additionally, sewage sludge from the waste water treatment process may be released into the environment.

Another direct route for PhACs to enter water bodies is through surface runoff. In this context, livestock manure is commonly used as an organic fertilizer to improve soil quality by providing essential nutrients and chemicals such as nitrogen and phosphorus, thereby enhancing crop yields. However, livestock manure contains PhACs or their corresponding metabolites, which can be easily absorbed into the soil or enter groundwater and surface water through runoff [10]. Thus, the application of organic fertilizers to agricultural land introduces PhACs into the soil, which may eventually leach into nearby surface water. The accumulation of PhACs in surface water systems poses serious environmental and health concerns, as they may enter rivers and lakes that are used for drinking water, thereby posing potential risks to human health [11].

It is also noteworthy that landfilling is one of the simplest and most common waste disposal practices. In this context, a significant pathway for PhACs to enter the environment is through groundwater. After long retention periods, PhACs from biosolids can migrate into groundwater through landfill leachate. Several studies have indicated that landfill leachate contains trace amounts of PhACs [12]. The variety and concentration of PhACs found in leachate are directly related to the volume of biosolids disposed, which is closely proportional to population density. PhACs released into groundwater can be easily absorbed by nearby plants and may eventually be transferred through the food chain, posing potential risks to ecosystems and human health.

3. Environmental impacts of pharmaceuticals

It has been pointed out that even at low concentrations, once PhACs enter water systems through any pathway, they begin to exert their effects on aquatic ecosystems. This can be rationalized by the fact that PhACs are specially designed and produced to be intrinsically bioactive, targeting pathogenic living cells and microorganisms in humans and animals. Therefore, their unintended presence in aquatic environments, water systems, or even in drinking water sources is a significant cause for concern. Although the ecological impacts, occurrence, fate, and long-term toxicity of the vast majority of PhACs in the environment remain unassessed, several PhACs have been shown to cause adverse effects on aquatic ecosystems. The most vulnerable organisms are aquatic wildlife, which are constantly exposed to PhACs throughout their entire lifecycle in the contaminated water. Furthermore, many PhACs are persistent in the aquatic environment, and their contamination and impacts are worsened by their continuous input into water systems. For instance, PhACs have been reported to cause direct toxic effects on reproduction, physiological functions, and behavior in organisms such as fish, water fleas, and duckweed, threatening their survival [2, 3].

The ecotoxicological effects of PhACs, especially antibiotics and non-steroidal anti-inflammatory pharmaceutical drugs – which are designed to kill or inhibit living organisms – on aquatic organisms include cytotoxicity, genotoxic damage, high oxidative cell stress, biochemical disruption, endocrine disruption, behavioral changes, and even population collapse. In general, biochemical disruption and endocrine disruption often lead to hormonal imbalances, causing increased lipid peroxidation, elevated antioxidant activity, and damage to gills, livers, and muscle tissue. These disruptions also contribute to reduced reproductive rates and the feminization of male fish.

The bioaccumulation and biomagnification of PhACs and their metabolites in aquatic environments are also linked to the increasing emergence of bacteria resistant to current pharmaceuticals. In this context, bacteria exposed to PhACs may develop antimicrobial resistance genes, which can be transferred to other animals and human pathogens. These resistant bacteria can enter the food chain through various environmental media, such as soil, water, air, sediment, or diets consumed at higher trophic levels. The effects of unintended exposure to PhACs, particularly those released from various pharmaceutical residues, on human health are still under investigation. Although the long-term impacts have not been fully confirmed, multiple studies suggest potential risks. For instance, it has been pointed out that certain pharmaceuticals have been associated with breast and testicular cancers, reduced fertility, reproductive disorders, and antibiotic-resistant infections. Moreover, the cumulative effects of chronic low-level exposure to PhACs can affect immune responses and the natural protection of antibodies, increasing human susceptibility to diseases [13].

4. Detection of PhACs

Accurate detection of PhACs and their metabolites is critical for identifying their presence in waste water and water systems, particularly as part of early environmental screening and warning systems. Extensive research has been devoted to quantifying and identifying these contaminants, often at the detection limit in the range of nanograms per liter to micrograms per liter. However, because PhACs and their metabolites are typically present at extremely low concentrations in the environment, their detection and identification remain significant analytical challenges.

The most crucial step in the detection and identification of PhACs involves collecting large volumes of water and subsequently concentrating the samples, often through solvent evaporation, which is a time-consuming process. Due to the low concentrations of these compounds, highly sensitive sensing techniques with exceptionally low detection limits are required. Another crucial step is data analysis, which is particularly complex, as real waste water often contains multiple PhACs and unknown metabolites, making interpretation more challenging.

The challenge is further amplified during the identification process, as the data obtained from measurements are often overlapping. This is primarily due to structural and physicochemical similarities, as well as the low separation efficiencies of PhACs and their metabolites. As a result, analytical systems with high resolution and high separation efficiencies are essential to differentiate individual PhACs. Among the most promising and robust analytical techniques for accurate detection are chromatography, spectroscopy, mass-spectrometry, electrochemical, optical, immunoassay, and microbial biosensors, as summarized in **Figure 2**.

Chromatography techniques separate complex mixtures into individual components based on their different interactions with the stationary and mobile phases. Each component travels at a different speed, allowing for the separation, detection, and quantification of PhACs and their metabolites. Common chromatography techniques include gas chromatography (GC) and high-performance liquid chromatography (HPLC). GC can detect volatile and

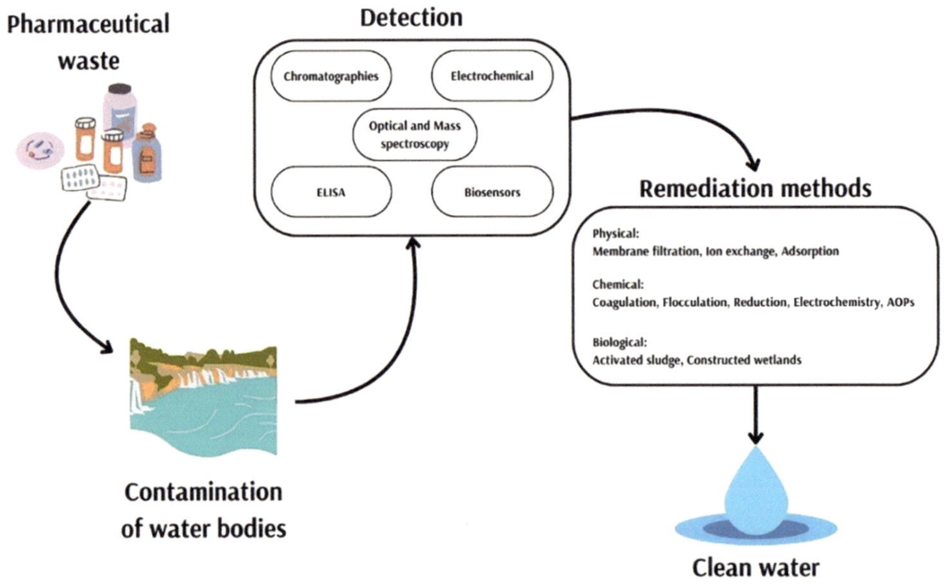

Figure 2.
Detection and identification of trace amounts of PhACs in water systems, along with promising techniques for their removal from contaminated water bodies.

semi-volatile PhACs. In this method, the liquid analytes are vaporized, and the resulting gas mixture is transported through a column by a mobile phase consisting of inert helium (He) or nitrogen (N_2) gas under high pressure. Separation of volatile and semi-volatile PhACs and their metabolites is based on their respective boiling points and affinity for the stationary phase. In HPLC, a liquid mobile phase carries the sample through columns under high pressure, separating individual constituents based on their different masses and affinities for the stationary phase. HPLC is suitable for identifying polar, non-volatile, and thermally unstable PhACs. Before analysis, analytes are commonly purified and concentrated through solid-phase extraction. Compared to GC, HPLC provides more accurate chemical identification and is ideal for the quantitative detection of most PhACs, as a majority of them are polar, non-volatile, and prone to decomposition at high temperatures [14].

Several spectroscopic methods, including ultraviolet–visible (UV-vis) spectroscopy, Fourier-transform infrared spectroscopy, Raman spectroscopy, Raman hyperspectral imaging, X-ray fluorescence, and diffuse reflectance spectroscopy, are widely utilized to identify the molecular structures of PhACs, mainly due to their rapid, non-destructive nature and simple sample preparation. These methods enable accurate identification and quantification of PhACs and their unknown metabolites, even in complex mixtures. In this context, the measured spectra, which may contain overlapping signals from multiple PhACs, are deconvoluted using spectral databases of unique standard compounds. Spectroscopic methods are not only promising for the identification and quantification of PhACs and their metabolites but also for tracing PhACs in complex waste water [15].

Mass spectrometry is another powerful method for identifying and quantifying PhACs in water based on their mass-to-charge ratios and fragmentation patterns, which allow the determination of chemical structures and quantification of their trace amounts. Although mass spectrometry is a powerful detection method, it requires clean samples to perform optimally, so it is often coupled with chromatographic methods to purify the analytes before analysis. For complex waste water analysis, where compounds with the same mass may be present, tandem mass spectrometry is commonly applied to further identify and quantify target compounds that may share identical masses. The high selectivity, sensitivity, and structural confirmation ability of tandem mass-spectrometry make this technique the gold standard for trace detection of PhACs, especially when paired with chromatographic techniques.

PhACs are also detected using immunoassay techniques, with the most common method being enzyme-linked immunosorbent assay (ELISA). In this assay, antibodies are specifically designed to capture and bind PhACs as the target analytes. A second antibody, attached to an enzyme marker, is then introduced to sandwich the target analytes, after which any unbound PhACs and antibodies are removed. The enzyme linked to the antibody catalyzes a chemical reaction and produces a color with an intensity that indicates the concentration of each PhAC. The strength of this technique lies in its capability for rapid analysis of multiple PhACs simultaneously at low cost, making it viable for screening and identifying large numbers of contaminants. However, ELISA can sometimes give false results due to cross-reactivity with molecules that have very similar structures and binding affinities to the antibodies as the target molecules [16].

Electrochemical analysis, using either non-fabricated or fabricated electrodes, is also effective for detecting PhACs. The performance of this method can be optimized by adjusting several detection parameters. Most importantly, the sensitivity and selectivity of electrochemical analysis are enhanced by incorporating metallic nanoparticles and carbon-based nanomaterials into the electrodes, which can improve conductivity, stability, selectivity, and the detectable current for biosensing applications. Electrochemical immunoassays are capable of sensitive PhACs detection. For instance, an electrochemical immunoassay combined with square-wave voltammetry can detect β -lactam antibiotics, including ampicillin, amoxicillin, cefotaxime, and dicloxacillin, with exceptionally low detection limits [17]. Electrochemical methods have also been utilized to simultaneously detect various classes of PhACs, such as residues of non-steroidal and steroidal anti-inflammatory drugs.

Optical and electrochemical biosensors employing metal-immobilized layered double hydroxides are also of interest for detecting various PhACs due to their high adsorption capacity, selectivity, and stability. The performance of these systems can be further optimized through chemical modification of the bioreceptors, such as enzymes, aptamers, or yeasts, which are essential biomolecules acting as recognition agents for analytes and are important to be integrated with optical measurement systems [18]. Additionally, the incorporation of membrane-based nanocomposites in electrochemical biosensors has been developed for the rapid detection of antiretroviral residues in waste water. These optical and electrochemical biosensors offer several advantages in terms of speed, accuracy, and sensitivity, enabling real-time detection of multiple PhACs in waste water.

Microbial biosensors employing common microorganisms, such as bacteria, viruses, microalgae, and fungi, have also been used to detect PhACs in waste water. This method has already been reported for detecting heavy metals and herbicides in the environment and is gaining interest for applications in controlling water quality. The basic principle of these sensors involves the use of microorganisms to detect the presence of specific contaminants, including foodborne pathogens. Microbial biosensors offer several advantages, such as simplicity, stability, sensitivity, and the potential for continuous monitoring. The sensitivity of biosensors can be improved by incorporating noble metallic nanomaterials. However, this method has several notable drawbacks, including high cost, microbial genetic modification, and reliance on exogenous substrates.

Recent advances in machine learning and image recognition have opened up new possibilities for detecting and identifying PhACs based on their physical characteristics, such as color, particle shape, particle size, and trajectories in waste water [19]. Although this method is still in its early stages, the development of comprehensive databases and neural networks holds promise for the efficient and accurate identification, quantification, and tracking of the trajectories of PhACs in waste water through image analysis, as well as for preventing medication misuse.

5. Remediation technologies

Considering the potential adverse effects, particularly the continuous increase in diverse PhACs and their metabolites in aquatic environments, various promising methods have been developed to address the challenges of removing these contaminants from water systems. These recently developed methods differ significantly from conventional water treatment technologies typically applied in WWTPs. As summarized in **Figure 2**, the advanced remediation methods that have been successfully shown to have potential for removing PhACs from waste water are broadly classified into physical, chemical, and biological techniques, as well as their combined and integrated approaches.

Common physical methods include membrane filtration, ion exchange, and adsorption. Chemical methods encompass oxidation, photo-oxidation, advanced oxidation processes (AOPs), sodium borohydride (NaBH4) reduction, and photosensitized oxidation. Biological methods include activated sludge systems and constructed wetlands. Each of these promising techniques offers distinct advantages, such as ease of use, simplicity, affordability, and high efficacy, although each also has disadvantages and certain limitations.

5.1 Physical approaches

5.1.1 Membrane filtration

Membrane filtration involves pumping waste water through semipermeable membranes with specific pore sizes. The separation of PhACs is simply based on molecular size exclusion. The process can be categorized based on pore size into microfiltration, ultrafiltration, nanofiltration, and reverse osmosis, each offering different levels of selectivity, effectiveness, and removal efficiency. Membrane filtration offers several advantages, including: (1) it does not rely on phase changes, making it more energy-efficient than distillation, (2) it results in minimal accumulation on membrane surfaces, allowing for continuous operation, and (3) its efficiency can be improved by incorporating chemical additives into the semipermeable membranes.

Although membrane filtration shows great potential for efficiently separating specific PhACs, several challenges remain, especially regarding performance, operational stability, and scalability. For example, laboratory-scale studies have shown that membrane filtration can achieve PhAC removal rates in the range of 10%–100%, depending on molecular size, pH of the medium, and the initial concentration of the contaminants. However, recently developed pilot-scale membrane filtration systems have reported maximum removal rates of approximately 56%–60% [20]. It is also important to note that membrane filtration requires the pre-treatment of waste water to improve separation efficiency. Due to issues such as membrane fouling or degradation, the permeable membranes need to be replaced regularly, significantly increasing operational costs. Furthermore, this method does not degrade or destroy PhACs but merely separates the contaminants, potentially creating secondary waste disposal issues. The high installation and maintenance costs, along with the generation of considerable amounts of toxic by-products, remain the key drawbacks of this technique.

5.1.2 Ion exchange

Ion exchange is a unique method for removing dissolved PhACs from waste water. This technique relies on the ion exchange capabilities of specially designed resin beads that attract and bind to the oppositely charged functional groups of

PhACs through electrostatic interactions. The resin beads can be functionalized with either weak or strong acid cations or weak or strong base anions, allowing the selective removal of a wide variety of PhACs. The ionizable functional groups attached to the resin beads are crucial in determining their binding capacity and selectivity for specific PhACs in waste water treatment systems. For example, strong acid resin beads functionalized with sulfonic acid (HSO_3^-) groups can effectively remove most cationic PhACs from water by exchanging them with hydrogen ions. Since this exchange reaction is reversible, the resin can be regenerated by treating it with an excessively strong acid solution, allowing for repeated use.

Ion exchange offers several advantages, including high removal efficiency for charged PhACs dissolved in water, applicability to a wide range of emerging contaminants, and the ability to customize resin beads for the selective removal of specific target PhACs. Therefore, this method plays a significant role in improving water quality and enabling the regeneration and recovery of PhACs, particularly those released from pharmaceutical industrial effluents. However, this technique also presents several limitations. These include the frequent replacement of resin beads to maintain high removal efficiency, high operational and installation costs, and the requirement for pretreatment of waste water to remove hardness ions such as magnesium, potassium, and calcium, which can interfere with the resin's performance. Additionally, the complex preparation of resin beads and the potential environmental issues associated with the disposal of by-products must be considered. Therefore, careful planning and evaluation are essential for the sustainable and effective application of ion exchange in waste water treatment.

5.1.3 Adsorption

Adsorption is a process in which PhACs dissolved in water are removed by adhering to the surfaces of solid materials, known as adsorbents, either in batch or column modes. The adsorption process occurs at the solid–liquid interface and is driven by Fick's first diffusion law, which governs the transfer of contaminants from the bulk solution to the interfacial layer of the adsorbent surface. Adsorption can be defined as the immobilization and accumulation of PhACs on the adsorbent surfaces, allowing them to be effectively separated from water. The reverse process, in which adsorbed PhACs are released back into the solution, is called desorption. The adsorption mechanism is associated with the immobilization of PhACs on the adsorbent surfaces, stabilized by hydrogen bonding and electrostatic interactions between the contaminants and the functional groups on the interfacial layer of the adsorbent. This technique is widely known for its versatility, simplicity, and high efficiency in removing toxic organic compounds from water [21]. Its versatility is reflected in the wide variety of available adsorbents. The effectiveness of adsorption in removing emerging contaminants largely depends on the physicochemical characteristics of the adsorbents, such as surface area, pore structure, surface morphology, net surface charge, and the presence of functional groups. In other words, these surface characteristics determine how effectively an adsorbent can attract and hold contaminants. Based on these fundamental principles, various adsorbent materials with large surface area, large pore volume, rough surface morphology, and abundant functional groups have been extensively investigated.

Activated carbon, clay minerals, agricultural waste-derived materials, and biopolymers have been identified as the most promising adsorbents. Among them, activated carbon is the most extensively utilized synthetic adsorbent due to its large surface area and exceptional adsorption efficiency for a wide range of contaminants, including volatile organic compounds, pesticides, PhACs, and heavy metals. It is also highly porous and effective in removing both liquid and gaseous contaminants. Clay minerals, such as bentonite and kaolinite, are commonly used because they are readily available, abundant, cost-effective, and have high adsorption capacity, especially for heavy metals, organic contaminants, and PhACs due to their high cation exchange capacity [22, 23]. Additionally, zeolites, which are naturally occurring aluminosilicates, exhibit well-defined molecular sieving properties and are effective at eliminating toxic molecules from water.

Agricultural waste, such as plant leaves, fruit peels, nut shells, and crop husks, is a sustainable and low-cost source of natural adsorbents. It is naturally abundant and rich in functional groups, such as carboxylic acid (COOH), amine (NH_2), sulfonic acid (SO_3H), carboxyl (CO), carbonic acid (CO_3H), and hydroxyl (OH) groups, which readily interact with PhACs [24]. Agricultural waste is frequently converted into biochar through pyrolysis, producing highly porous carbon-rich materials suitable for adsorbing organic pollutants. Biopolymers, especially cellulose, chitosan, and low-methoxy pectin, are another class of adsorbents and play a crucial role in removing numerous contaminants, including dyes, heavy metals, organic compounds, and PhACs from waste water. Cellulose is derived from plant leaf fibers, pectin from fruit peels, and chitosan converted from chitin, a substance mainly obtained from the exoskeletons of crustaceans, such as shrimp, crab, and lobster. Their natural abundance and versatility make biopolymers an excellent choice as adsorbents in water treatment applications. The unique structures of biopolymers, which comprise abundant functional groups, such as COOH, NH_2, $COOCH_3$, and OH, bound to glucose, glucosamine, and galacturonic units of cellulose, chitosan, and low-methoxy pectin, respectively, offer numerous active binding sites [25].

Each type of adsorbent has distinct benefits and characteristics, but activated carbon, clay minerals, agricultural waste, and biopolymers have gained significant attention due to their physical and chemical stability, biodegradability, and biocompatibility, which fulfill the requirements for good adsorbents. With these properties, these adsorbents can degrade naturally into harmless by-products, making them environmentally friendly adsorbents without generating secondary contaminants. Their natural abundance aligns with the principles of environmental sustainability and economic feasibility. Moreover, versatile chemical modifications of these adsorbents allow various derivative materials to remove PhACs.

It should be noted that synthetic materials, such as polymeric resins and metal–organic frameworks, have also been developed for specialized applications, particularly for the removal of PhACs, dyes, pesticide residues, and heavy metals. Although these materials are designed for high selectivity and efficiency, often exceeding natural adsorbents in removing certain contaminants, they are less cost-effective and less sustainable.

In addition to surface characteristics, key factors influencing the adsorption capacity of PhACs are operational conditions, such as the pH of the medium, which affect surface charge and adsorbent–PhAC interactions, and the molecular properties, such as charge distribution, polarity, molecular size, and conformational flexibility of the PhACs themselves, . These molecular properties determine how PhACs are arranged on the adsorbent surfaces. For instance, non-planar PhACs typically form a monolayer without significant intermolecular interactions, whereas planar compounds can form multilayer stacking through π–π interactions, thereby increasing overall adsorption capacity [25]. A deeper understanding of the adsorption mechanisms can be achieved by analyzing experimental data using various isotherm, kinetic, and diffusion models. These insights are essential for optimizing adsorbent design to enable more efficient removal of PhACs from waste water.

It is interesting to note that, in practical scenarios, typical waste water contains multiple PhACs. In most cases, simultaneous adsorption of multiple contaminants should be competitive, as they compete for the same functional groups on the adsorbent surfaces. This competition for binding sites often reduces the adsorption capacity for individual compounds. However, synergistic effects have also been observed, for example, in the simultaneous adsorption of Methylene Blue and Acid Red-1, where oppositely charged organic compounds exhibit attractive electrostatic interactions and stack to form multilayers on the adsorbent surfaces, thereby enhancing the overall adsorption capacity [26]. Similar competitive and synergistic phenomena are believed to occur during the concurrent adsorption of multiple PhACs.

As mentioned above, one of the main advantages of adsorption is its simple procedure. The process requires no sophisticated equipment or technical expertise, involves low operational costs, and allows the use of a wide range of optional adsorbents. Adsorption is also highly effective at low concentrations of PhACs, making it ideal for treating trace levels of the emerging contaminants in water. Moreover, many adsorbents can often be regenerated and reused after treatment, further enhancing cost-effectiveness and minimizing environmental impact.

5.2 Chemical approaches

5.2.1 Coagulation–flocculation

Coagulation and flocculation are complementary processes widely applied in waste water treatment, as this combination effectively removes large colloidal or suspended particles of disperse dyes, coloring agents, and PhACs. Coagulation is a chemical process in which additional coagulants, such as aluminum sulfate ($Al_2(SO_4)_3$), ferric sulfate ($Fe_2(SO_4)_3$), and calcium hydroxide ($Ca(OH)_2$), are added to destabilize suspended or colloidal particles in contaminated water. These coagulants dissociate into cations and anions, which react with the counter ions of suspended contaminants to form neutral salts. These small suspended neutral salt formations aggregate to form micro-flocs. Flocculation then induces the aggregation of these micro-flocs into larger and stable flocs, which can settle more easily and be removed through filtration or sedimentation.

Flocculation is typically classified into micro-flocculation (or perikinetic flocculation) and macro-flocculation (orthokinetic flocculation) based on floc size. In micro-flocculation, micrometer-scale flocs aggregate due to Brownian motion. In macro-flocculation, larger flocs are formed due to external forces such as stirring. In this case, rapid stirring is necessary at this stage to disperse the coagulants and to promote collisions between floc particles for effective aggregation. However, the effectiveness of the coagulation–flocculation process can be hindered by electrostatic repulsion between similarly charged ions, which inhibits the aggregation of small flocs to form larger aggregates. As a result, this method is generally ineffective for small or highly soluble PhACs, which remain dissolved and do not readily form flocs.

5.2.2 Reduction by sodium borohydride

Sodium borohydride ($NaBH_4$) is a well-known strong reducing agent. It reduces the carbonyl groups of organic compounds, leading to their breakdown into smaller molecules. This method has been successfully applied to reduce effluents containing coloring agents, particularly azo dyes, in waste water. The presence of metal oxides, such as ZnO, SiO_2, and Fe_3O_4 nanoparticles, or noble metals, such as gold (Au) or silver (Ag) nanoparticles, typically catalyzes the reduction process.

The reduction of organic contaminants relies on the dissociation of $NaBH_4$ into sodium and borohydride ions. The reaction mechanism is then initiated when the borohydride anion donates electrons to the organic contaminants, triggering a redox process. This reaction is more efficient on the surface of catalyst nanoparticles. These catalysts serve a dual function: they provide a large surface area that facilitates interactions between the organic molecules and borohydride ions, and they offer a feasible pathway for electron transfer, thereby accelerating the overall redox reaction [27].

The efficiency of sodium borohydride in reducing organic contaminants can be optimized by controlling several operational parameters, including the pH of the medium, temperature, catalyst dosage, and the initial concentration of the emerging contaminants in the waste water. The progress of the reduction reaction can be monitored using UV-vis spectrometry, which tracks the decrease in absorbance associated with the concentration of the organic contaminants. However, the effectiveness of this method can be affected by the presence and nature of electrolytes in the waste water, which may reduce the available active sites on the catalyst surfaces and lower the overall reaction rate. Despite this limitation, sodium borohydride reduction remains an efficient method for degrading organic contaminants. Given its success in such applications, this method is considered promising for the removal of PhACs from waste water.

5.2.3 Electrochemistry

The electrochemical process for degrading PhACs in waste water has garnered significant interest, as it is considered to be an environmentally friendly approach. This method is robust and applicable for the degradation of a large variety of organic contaminants, such as aromatics, polymers, pesticides, and PhACs. The electrochemical process is typically performed in a batch reactor equipped with electrodes and a supporting electrolyte solution. By applying electrical energy, redox reactions are initiated, leading to the degradation of contaminants in the water solution into building block compounds such as carbon dioxide, carbon monoxide, hydrogen, nitrogen, and oxygen, which are harmless to the environment. The degradation mechanism of organic contaminants begins with ring-opening processes, followed by further molecular breakdown through dehydration, dehydroxylation, dealkylation, and bond breaking. Therefore, the electrochemical process is particularly effective for treating waste water containing recalcitrant organic contaminants on site, making this method a promising candidate for advanced waste water treatments.

The main challenges associated with the electrochemical process are its energy efficiency and overall efficacy in terms of energy consumption. To address these issues, various anode materials with high electrocatalytic activity, strong degradation capacity, and corrosion resistance have been intensively studied to achieve high oxidation rates. Among these promising materials, boron-doped diamond anodes and stainless-steel cathodes have emerged as the most efficient configuration for electrochemical reactors [28]. Several supporting electrolytes, such as H_2SO_4, Na_2SO_4, and NaCl, have been shown to enhance degradation efficiency, with H_2SO_4 exhibiting the highest performance due to the shortest half-life of the generated sulfate radicals.

The rate-limiting steps in the electrochemical process include the applied current density and the mass transfer of organic contaminants from the bulk solution to the anode surface. Overall operational costs, such as electrode supply and replacement, energy consumption, and reactor cleaning and maintenance, are also important considerations. Notably, the installation and operational expenses, particularly electricity costs, remain the most crucial barriers limiting the potential scalability of this method for large-scale waste water treatment applications.

5.2.4 Advanced oxidation processes

One of the most interesting innovations for degrading PhACs in aqueous solution is AOPs. These processes generate reactive oxygen species (ROS), highly reactive molecules that contain at least a single unpaired electron, such as superoxide ($^{\bullet}O_2^-$), hydroxyl ($^{\bullet}OH$), hydroperoxyl ($^{\bullet}HO_2$), singlet oxygen ($^1O_2$), ozone (O_3), and hydroperoxide anion (HO_2^-). These ROS are primarily responsible for oxidizing and breaking down PhACs. AOPs are considered efficient and environmentally friendly methods for treating refractory PhACs and their metabolites [29].

Various AOPs have been developed as promising alternatives for the rapid degradation and elimination of PhACs in waste water, converting them into biodegradable and environmentally benign products. These include UV–hydrogen peroxide (UV/H_2O_2), Fenton, photo–Fenton, electro–Fenton, ozone-based processes, organic photoredox catalysis, and photocatalysis. Numerous studies have particularly focused on the degradation of PhACs using AOPs [30–34]. This section provides an overview of the fundamental principles, advantages, disadvantages, and degradation mechanisms of PhACs in aqueous solution using each AOP.

The UV/H_2O_2 treatment utilizes UV light to initiate the photolytic dissociation of H_2O_2, producing $^{\bullet}OH$ radicals in aqueous solution. With a standard reduction potential of +2.80 V, higher than that of most oxidizing agents, the $^{\bullet}OH$ radical is a strong oxidant, capable of oxidizing and degrading a wide variety of PhACs. However, due to its short lifetime in water and the diffusion-limited nature of its reactions, the actual degradation rates can be relatively low, particularly at trace concentrations of PhACs. Additionally, high concentrations of H_2O_2 are often required, making this method less cost-effective [30, 31].

The Fenton reaction is based on the Haber–Weiss mechanism, in which ferrous (Fe^{2+}) ions are oxidized by H_2O_2 to produce $^{\bullet}OH$ radicals, with ferric (Fe^{3+}) ions generated as a by-product. The catalytic degradation rate of PhACs is limited by the rate of $^{\bullet}OH$ radical formation, which is governed by the concentration of Fe^{2+} and the amount of H_2O_2. When UV light is introduced, the photolytic dissociation of H_2O_2 is accelerated, leading to the so-called photo–Fenton process, which degrades organic contaminants more efficiently than the conventional Fenton reaction. The Fenton reaction can also be enhanced by applying an electric field. In an electro-Fenton cell, the generation of $^{\bullet}OH$ radicals is positively related to the applied electric field. It should be noted that Fenton, photo-Fenton, and electro-Fenton processes are relatively simple and require less technological complexity and equipment. However, if Fe^{3+} by-products are not converted back to Fe^{2+}, they can inhibit the Fenton reaction and cause secondary iron pollution. Another drawback is the requirement of acidic (pH 3) conditions, which necessitates the addition of strong acids and can lead to further environmental concerns.

Ozonation has been successfully established for the treatment of PhACs in waste water [33, 34]. With a reduction potential of +2.07 V, ozone is a powerful oxidizing agent capable of oxidizing and degrading PhACs either directly, through electrophilic attack by transferring one of its oxygen atoms to the emerging contaminants, or indirectly, by generating $^{\bullet}OH$ radicals in water. Ozone is typically produced by passing plasma through oxygen gas and then introduced into waste water as microbubbles, which enhance the generation of $^{\bullet}OH$ radicals on the gas-liquid interface. The addition of H_2O_2 or UV light can further accelerate the degradation process of PhACs [35]. Microbubble ozonation is particularly effective and results in minimal harmful by-products. However, its application is limited by the low solubility of ozone in water, high energy consumption, and the potential formation of genotoxic by-products.

Organic photoredox catalysts harness visible light to convert excitation energy into singlet oxygen ($^1O_2$), a selective oxidant with a reduction potential of +1.48 V. This allows selective oxidation and degradation of organic pollutants in waste water under mild conditions [36]. Several negatively charged organic dyes, such as Eosin Y and Rose Bengal, have been widely studied as cost-effective photosensitizing compounds for organic photoredox catalysts. Organic photoredox catalysis is considered an eco-friendly treatment method, but the detailed mechanisms remain an active area of research. This method also faces several challenges, such as limited light penetration in turbid waste water and potential photobleaching of the organic photoredox catalysts, both of which reduce the overall photocatalytic efficiency.

Photocatalysis utilizes light energy, allowing electrons to be excited from the valence band to the conduction band of a photocatalyst, leaving holes in the valence band. The photocatalytic degradation of organic pollutants occurs on the photocatalyst surface when the photogenerated charge carriers migrate to the surfaces of photocatalysts and undergo redox reactions, producing ROS such as $^{\bullet}O_2^-$ and $^{\bullet}OH$ radicals. Typical photocatalysts with high intrinsic photoactivity are metal oxides, perovskite-type metal oxides, and metal sulfide nanoparticles, such as TiO_2, ZnO, CeO_2, Mn_3O_4, $SrTiO_3$, $BiVO_4$, Bi_2WO_6, CdS, and ZnS. These photoactive metal oxides are even more fascinating because their morphologies, nanostructures, and bandgap energies can be engineered to optimize their performance.

Photocatalysis employing TiO_2 and $SrTiO_3$ nanoparticles as photocatalysts has been utilized to degrade various PhACs in artificial waste water [30–32]. Moreover, TiO_2-based photocatalysis has also been implemented in pilot projects for treating PhACs in real waste water in Poland and the United States. Detailed studies have pointed out that the photocatalytic degradation kinetics and rates of PhACs depend on several operational parameters, such as the photocatalyst dosage, pH of the medium, and the addition of H_2O_2. More importantly, intrinsic properties of PhACs, such as their chemical and conformational structures, attached functional groups, and charge delocalization, play a key role in determining diffusion, immobilization, surface coverage on the photocatalyst, and photoreaction coefficient [37].

Photocatalysis offers advantages such as complete mineralization, versatility, reusability, and cost-effectiveness of photocatalysts. However, the photoexcited electrons often suffer from recombination, which limits the redox reactions on the photocatalyst surfaces, reduces the generation of $^{\bullet}O_2^-$ and $^{\bullet}OH$ radicals, and thereby decreases photocatalytic activity. Additionally, the photocatalytic degradation pathways of PhACs are also of concern due to the possible formation of intermediates and by-products, which may be more harmful to the environment and aquatic organisms than the parent PhACs.

5.3 Biological approaches

5.3.1 Activated sludge

Activated sludge is a widely used biological process in waste water treatment. This method involves a bacterial biomass suspension, commonly referred to as activated sludge, containing bacteria and protozoa. When the activated sludge is introduced into waste water, these organisms decompose organic compounds through metabolic processes. The activated sludge process requires controlled aeration, agitation, and separation. After aeration and agitation, the waste water is collected in a clarifier tank, allowing the activated sludge to settle. The settled sludge and the remaining water are then either discharged or reused. To maintain the efficiency of the activated sludge process, the sludge is returned to the aeration tanks to conserve the microbial population. In addition, aeration time, agitation, and dissolved oxygen levels in the aeration tank must be carefully regulated.

The activated sludge process has shown high efficiency in removing several PhACs, especially antibiotics [38]. Several pilot-scale projects have also combined the activated sludge process with biofilms. Although this combined approach has revealed near-complete removal of certain PhACs within 24 hours of treatment, it still has some disadvantages, including high oxygen consumption, large installation space, high amounts of waste sludge, vulnerability to external disturbances, and operational costs. Another major concern is the potential for bacterial resistance. Prolonged exposure to PhACs can lead to the development of resistant bacterial strains, posing challenges for sludge disposal, increasing the risk of secondary pollution, and complicating microbial cycling in the treatment system.

5.3.2 Constructed wetlands

Constructed wetlands are designed and engineered systems utilizing natural processes involving existing soils, wetland vegetation, and related microorganisms, such as bacteria, yeasts, fungi, protozoa, and filamentous algae, in a controlled environment. A wetland is a unique ecosystem based on an area flooded with water, enabling the growth of aquatic vegetation. The design of constructed wetlands depends on the surrounding natural conditions. Typically, a constructed wetland includes surface and subsurface water flow, a shallow artificial basin filled with aquatic emergent, submerged, free-floating, or floating leafy plants, and a defined water flow path. As waste water passes through the constructed wetland, the plants absorb certain contaminants and create a microenvironment where microorganisms can grow and decompose PhACs. Therefore, constructed wetlands involve a synergistic combination of filtration, sedimentation, and adsorption processes by the soils, plants, and microorganisms, all of which contribute to the breakdown and removal of the emerging contaminants.

Constructed wetlands, as natural purification systems, have been widely applied for removing PhACs from a variety of waste water types. The advantages of this method include cost-effectiveness in construction, operation, and maintenance, feasibility, versatility, and water recycling. Additionally, the wetlands can serve as habitats for wildlife, supporting economic, ecological, and esthetic values, as the system can be integrated into landscape designs. It is worth highlighting that constructed wetlands have been extensively studied to optimize their performance by testing different low-cost substrates, designs, and operational parameters for the effective degradation and removal of PhACs [39].

Pilot-scale applications have shown promising results by adjusting several parameters, such as the need for aeration, concentrations of pollutants, and retention time in the hydraulic system. Despite these benefits, common challenges with this method include the requirement for large land areas, susceptibility to climate variations, decreased contaminant removal capability over time, and the susceptibility of aquatic life to PhACs that may induce antibiotic-resistant genes. This risk of promoting antibiotic-resistant genes remains a significant concern in aquatic environments, including wetland systems.

6. Future perspectives

With the continuous accumulation of PhACs in water systems, increasing research efforts have been devoted to mitigating their adverse effects on the environment, ecosystems, and human health. Currently, several promising remediation techniques, based on different physical, chemical, and biological approaches, each with distinct characteristic features, have been developed to treat various PhACs. However, these techniques still face several challenges, particularly regarding removal and degradation efficiency, optimization of operational parameters, cost-effectiveness, economic feasibility, toxicity, and risk assessment of by-products formed during and after contaminant removal or degradation.

Improving removal efficiency remains a key priority for physical approaches, such as membrane filtration, ion exchange, and adsorption. However, these approaches typically require further treatment of the separated PhACs to prevent secondary pollution, as the separated PhACs are often disposed of through incineration or landfilling. In the future, desorption and reuse of adsorbents and PhACs will become increasingly important to improve sustainability. In contrast, chemical approaches, such as coagulation–flocculation, chemical reduction, electrochemical treatment, and AOPs, focus on enhancing degradation efficiency to ensure the complete degradation and mineralization of PhACs. Meanwhile, biological approaches, including activated sludge and constructed wetlands, must address issues related to the formation of secondary pollutants, the development of bacterial resistance, and the susceptibility of aquatic life.

To address the abovementioned challenges, considering that some of the remediation techniques are flexible, combinatory treatments, combining multiple remediation techniques, either simultaneously or sequentially, represent a promising strategy. For example, several AOPs can be combined to improve the degradation efficiency of PhACs and their metabolites. Similarly, combining or integrating membrane filtration with biological treatments provides a comprehensive approach for effective contaminant removal. In this context, the membrane filtration serves as a physical barrier to separate suspended particulates, while microorganisms biodegrade dissolved contaminants into non-toxic by-products.

Other promising combined treatments are bio-photoelectrochemical, bio–Fenton, and photoelectrochemical processes. Bio–photoelectrochemistry integrates biological, photochemical, and electrochemical processes, utilizing sunlight, microorganisms, and catalysts to degrade PhACs. The bio–Fenton process combines Fenton chemistry and biological treatment to overcome their individual shortcomings. This sequential process begins with the Fenton reaction, which enhances the biodegradability of contaminants, thus facilitating subsequent biological degradation.

Photoelectrochemical techniques encompass various methods such as photoelectrocatalysis, photoelectron–Fenton, electro–peroxone, photoelectro–peroxone, photoanodic oxidation, and photocatalytic fuel cells. These approaches combine electrochemical-based AOPs, which are widely regarded as the most effective techniques for treating multiple PhACs in waste water, with light irradiation, Fenton chemistry, ozonation, photooxidation, or photocatalytic oxidation. These integrated remediation strategies offer some of the most effective solutions for the simultaneous treatment of multiple PhACs in waste water, aligning with broader environmental protection and conservation goals.

7. Conclusions

PhACs are continuously discharged into waste water, posing ecological and public health risks due to their ecotoxicity, biological activity, and adverse effects on both the environment and human health. The sources of PhACs, their transport pathways into water systems, their toxicity and environmental impacts, and their threats to human health have been extensively summarized. Even at low concentrations, PhACs can cause undesirable effects, particularly carcinogenic and mutagenic potentials, hormone disruption, behavioral disorders, population decline in aquatic organisms, and the development of antimicrobial resistance genes. Accurate detection methods capable of identifying and quantifying PhACs at trace levels in waste water and water systems play a crucial role as early environmental monitoring tools. Understanding the presence of these emerging contaminants raises awareness of the urgent need for their removal and degradation using appropriate remediation techniques.

Several remediation techniques, including membrane filtration, ion exchange, adsorption, coagulation–flocculation, chemical reduction, electrochemical treatment, AOPs, activated sludge systems, and constructed wetlands, have been discussed in detail along with their advantages and limitations. These techniques are efficient in removing and degrading PhACs and show strong potential for large-scale application in waste water treatment systems. Despite their effectiveness, gaps remain, particularly in the integration of treatment processes. Future efforts should focus on developing integrated, cost-effective remediation technologies, supported by global collaboration for monitoring, data sharing, and coordination. Such collaborative approaches are essential to protect water bodies and ecosystems from the growing threat posed by PhACs as emerging environmental contaminants.

Conflict of interest

The authors declare no conflict of interest.

Author details

Harry Lik Hock Lau[1], Rusydi R. Sofian[1], Syahirah Nabilah Aedy Aewandy[1], Nur Diana Bazilah Awang Idris[1], Nur Aisyah Abdul Munir[1], Nur Nabaahah Roslan[1], Aisyah Farhanah Abdul Majid[1], Anwar Usman[1*] and Muhammad Nur[2]

1 Faculty of Science, Department of Chemistry, Universiti Brunei Darussalam, Bandar Seri Begawan, Brunei Darussalam

2 Center for Plasma Research, Integrated Laboratory, Universitas Diponegoro, Tembalang Campus, Semarang, Indonesia

*Address all correspondence to: anwar.usman@ubd.edu.bn

References

[1] Ortúzar M, Esterhuizen M, Olicón-Hernández DR, González-López J, Aranda E. Pharmaceutical pollution in aquatic environments: A concise review of environmental impacts and bioremediation systems. Frontiers in Microbiology. 2022;**13**:869332. DOI: 10.3389/fmicb.2022.869332

[2] Białk-Bielińska A, Grabarczyk Ł, Mulkiewicz E, Puckowski A, Stolte S, Stepnowski P. Mixture toxicity of six pharmaceuticals towards *Aliivibrio fischeri*, *Daphnia magna*, and *Lemna minor*. Environmental Science and Pollution Research. 2021;**29**:26977–26991. DOI: 10.1007/s11356-021-17928-y

[3] Montiel-Mora JR, Méndez-Rivera M, Ramírez-Morales D, Cambronero-Heinrichs JC, Rodríguez-Rodríguez CE. Toxicity of selected pharmaceuticals and their mixtures to the aquatic indicators *Daphnia magna* and *Aliivibrio fischeri*. Ecotoxicology. 2024;**33**:1047–1061. DOI: 10.1007/s10646-024-02798-0

[4] Naseem A, Alneghery LM, Al-Zharani M, Nasr FA, Jawad SS, Umer M, Sayyed R, Ilyas N. An insight into the impacts of pharmaceutical pollutants on the ecosystem and the potential role of bioremediation in mitigating pharmaceutical pollutants. International Journal of Pharmaceutics. 2025;**680**:125791. DOI: 10.1016/j.ijpharm.2025.125791

[5] Peña OIG, Zavala MÁL, Ruelas HC. Pharmaceuticals market, consumption trends and disease incidence are not driving the pharmaceutical research on water and waste water. International Journal of Environmental Research and Public Health. 2021;**18**:2532. DOI: 10.3390/ijerph18052532

[6] Rusu L, Grigoraș CG, Simion AI, Suceveanu EM, Șuteu D, Harja M. Application of saccharomyces cerevisiae/calcium alginate composite beads for cephalexin antibiotic biosorption from aqueous solutions. Materials. 2021;**14**:4728. DOI: 10.3390/MA14164728

[7] Kumar V, Lakkaboyana SK, Sharma N, Chakraborty P, Umesh M, Pasrija R, Thomas J, Kalebar VU, Jayaraj I, Awasthi MK, Das T, Oladipo AA, Barcelo D, Dumee LF. A critical assessment of technical advances in pharmaceutical removal from waste water – A critical review. Case Studies in Chemical and Environmental Engineering. 2023;**8**:100363. DOI: 10.1016/j.cscee.2023.100363

[8] Pech-Rodríguez WJ, Meléndez-González PC, Hernández-López JM, Suarez-Velázquez GG, Sarabia-Castillo CR, Calles-Arriaga CA. Pharmaceutical waste water and sludge valorization: A review on innovative strategies for energy recovery and waste treatment. Energies. 2024;**17**:5043. DOI: 10.3390/en17205043

[9] Bound JP, Voulvoulis N. Household disposal of pharmaceuticals as a pathway for aquatic contamination in the United Kingdom. Environmental Health Perspectives. 2005;**113**:1705–1711. DOI: 10.1289/ehp.8315

[10] Silva-Leal JA, Pérez-Vidal A, Torres-Lozada P. Effect of biosolids on the nitrogen and phosphorus contents

of soil used for sugarcane cultivation. Heliyon. 2021;7:e06360. DOI: 10.1016/j.heliyon.2021.e06360

[11] Pinto SM, Gomes IB. An overview of the impact of pharmaceuticals on aquatic microbial communities. Antibiotics. 2022;**11**:1700. DOI: 10.3390/antibiotics11121700

[12] Yu X, Lyu S, Zhao W, Guo C, Xu J, Sui Q. A picture of pharmaceutical pollution in landfill leachates: Occurrence, regional differences and influencing factors. Waste Management. 2024;**184**:20–27. DOI: 10.1016/j.wasman.2024.05.019

[13] Bacanlı MG. The two faces of antibiotics: An overview of the effects of antibiotic residues in foodstuffs. Archives of Toxicology. 2024;**98**: 1717–1725. DOI: 10.1007/s00204-024-03760-z

[14] Meng Y, Liu W, Liu X, Zhang J, Peng M, Zhang T. A review on analytical methods for pharmaceutical and personal care products and their transformation products. Journal of Environmental Sciences. 2021;**101**: 260–281. DOI: 10.1016/j.jes.2020.08.025

[15] Farrukh MA, editor. Advanced Aspects of Spectroscopy. Rijeka, Croatia: IntechOpen; 2012. 550 p. DOI: 10.5772/2757

[16] Jaria G, Calisto V, Otero M, Esteves VI. Monitoring pharmaceuticals in the aquatic environment using enzyme-linked immunosorbent assay (ELISA)—a practical overview. Analytical and Bioanalytical Chemistry. 2020;**412**:3983–4008. DOI: 10.1007/s00216-020-02509-8

[17] Pollap A, Kochana J. Electrochemical immunosensors for antibiotic detection. Biosensors. 2019;**9**:61. DOI: 10.3390/bios9020061

[18] Wang X, Zhou J, Wang H. Bioreceptors as the key components for electrochemical biosensing in medicine. Cell Reports Physical Science. 2024; **5**:101801. DOI: 10.1016/j.xcrp.2024.101801

[19] Kim S, Chae M, Lee J, Lee H. Advanced pharmaceutical recognition system based on deep learning for mobile medication identification. Applied Sciences. 2025;**15**:5644. DOI: 10.3390/app15105644

[20] Wang Q, Tang X, Zeng W, Wang F, Gong W, Chen J, Wang J, Li G, Liang H. Pilot-scale biological activated carbon filtration–ultrafiltration system for removing pharmaceutical and personal care products from river water. Water. 2022;**14**:367. DOI: 10.3390/w14030367

[21] Zaidi NAHM, Lim LBL, Priyantha N, Usman A. *Artocarpus odoratissimus* leaves as an eco-friendly adsorbent for the removal of toxic rhodamine B dye in aqueous solution: Equilibrium isotherm, kinetics, thermodynamics and regeneration studies. Arabian Journal for Science and Engineering. 2018;**43**:6011–6020. DOI: 10.1007/s13369-018-3228-9

[22] Majid AFA, Dewi R, Shahri NNM, Shahrin EWES, Kusrini E, Shamsuddin N, Lim JW, Thongratkaew S, Faungnawakij K, Usman A. Enhancing adsorption performance of alkali activated kaolinite in the removal of antibiotic rifampicin from aqueous solution. Colloids and Surfaces A: Physicochemical and Engineering Aspects. 2023;**676**:132209. DOI: 10.1016/j.colsurfa.2023.132209

[23] Majid AFA, Sofri NANH, Mahdzidi NS, Shamsuri N, Kusrini E, Thongratkaew S, Faungnawakij K, Nur M, Usman A. Kaolin-based eco-friendly adsorbents for the removal cephalexin antibiotic: Insights into kinetics, isotherm, mechanism, thermodynamic, and adsorbent regeneration. Journal of Dispersion Science and Technology. 2025. DOI: 10.1080/01932691.2025.2563809

[24] Shahrin EWES, Narudin NAH, Shahri NNM, Nur M, Lim JW, Bilad MR, Mahadi AH, Hobley J, Usman A. A comparative study of adsorption behavior of rifampicin, streptomycin, and ibuprofen contaminants from aqueous solutions onto chitosan: Dynamic interactions, kinetics, diffusions, and mechanisms. Emerging Contaminants. 2023;**9**:100199. DOI: 10.1016/j.emcon.2022.100199

[25] Shahrin EWES, Narudin NAH, Shahri NNM, Verinda SB, Nur M, Hobley J, Usman A. Adsorption behavior and dynamic interactions of anionic acid blue 25 on agricultural waste. Molecules. 2022;**27**:1718. DOI: 10.3390/molecules27051718

[26] Asbollah MA, Mahadi AH, Kusrini E, Usman A. Synergistic effect in concurrent removal of toxic methylene blue and acid red-1 dyes from aqueous solution by durian rind: Kinetics, isotherm, thermodynamics, and mechanism. International Journal of Phytoremediation. 2021;**23**:1432–1443. DOI: 10.1080/15226514.2021.1901851

[27] Al Rubai HF, Sultan MS, Abd GM. Enhancing the removal efficiency of azo dye using sodium borohydride in the presence of inorganic nano catalysts. Green Analytical Chemistry. 2025;**12**:100194. DOI: 10.1016/j.greeac.2024.100194

[28] Nashat M, Mossad M, El-Etriby HK, Alalm MG. Optimization of electrochemical activation of persulfate by BDD electrodes for rapid removal of sulfamethazine. Chemosphere. 2022;**286**:131579. DOI: 10.1016/j.chemosphere.2021.131579

[29] Roslan NN, Lau HLH, Suhaimi NAA, Shahri NNM, Verinda SB, Nur M, Lim JW, Usman A. Recent advances in advanced oxidation processes for degrading pharmaceuticals in waste water—A review. Catalysts. 2024;**14**:189. DOI: 10.3390/catal14030189

[30] Roslan NN, Lau HLH, Jali NDR, Yussof NAS, Nur M, Taha H, Kusrini E, Thongratkaew S, Kajornsak Faungnawakij K, Usman A. Photocatalytic degradation of cephalexin antibiotic on TiO2 nanoparticles: Insights from kinetics, thermodynamics, liquid chromatography-mass spectrometry, degradation pathways, and antibacterial activities. Reaction Kinetics, Mechanisms and Catalysis. 2025;**138**:1175–1196. DOI: 10.1007/s11144-025-02813-3

[31] Suhaimi NAA, Umar MKH, Lau HLH, Roslan NN, Lim JW, Hobley J, Nur M, Usman A. An insight into the photocatalytic degradation of the antibiotic rifampicin by titanium dioxide nanoparticles in aqueous solution under UV light irradiation. Reaction Kinetics, Mechanisms and Catalysis. 2024;**137**:1105–1123. DOI: 10.1007/s11144-023-02564-z

[32] Lau HLH, Yussof NAS, Roslan NN, Kusrini E, Prasetyo AB, Taha H, Thongratkaew S, Faungnawakij K, Nur M, Usman A. Photocatalytic degradation of cephalexin and rifampicin antibiotics on cubic SrTiO3 nanoparticles: Kinetics, rate limiting steps, thermodynamics, degradation

pathways, and bioactivites of photocatalytic degradation products. Reaction Kinetics, Mechanisms and Catalysis. 2025. DOI: 10.1007/s11144-025-02978-x

[33] Verinda SB, Amalia NF, Gunawan G, Farida H, Yulianto E, Usman A, Nur M Mitigation of emerging contaminants in hospital waste water: Ozone microbubbles as an innovative treatment approach. In: E3S Web of Conference. 2024;**503**: 01005. DOI: 10.1051/e3sconf/202450301005

[34] Verinda SB, Muniroh M, Yulianto E, Maharani N, Gunawan G, Amalia NF, Hobley J, Usman A, Nur M. Degradation of ciprofloxacin in aqueous solution using ozone microbubbles: Spectroscopic, kinetics, and antibacterial analysis. Heliyon. 2022;**8**: e10137. DOI: 10.1016/j.heliyon.2022.e10137

[35] Patel S, Mondal S, Majumder SK, Das P, Ghosh P. Treatment of a pharmaceutical industrial effluent by a hybrid process of advanced oxidation and adsorption. ACS Omega. 2020;**5**:32305–32317. DOI: 10.1021/acsomega.0c04139

[36] Fabregat V, Pagán JM. A green chemistry and energy- and cost-effective approach in innovative advanced oxidation processes through photoactive microgels for sustainable applications. Sustainability. 2025;**17**(5):2331. DOI: 10.3390/su17052331

[37] Suhaimi NAA, Kong CPY, Shahri NNM, Nur M, Hobley J, Usman A. Dynamics of diffusion- and immobilization-limited photocatalytic degradation of dyes by metal oxide nanoparticles in binary or ternary solutions. Catalyst. 2022;**12**:1254. DOI: 10.3390/catal12101254

[38] Quintelas C, Mesquita DP, Torres AM, Costa I, Ferreira EC. Degradation of widespread pharmaceuticals by activated sludge: Kinetic study, toxicity assessment, and comparison with adsorption processes. Journal of Water Process Engineering. 2020;**33**:101061

[39] Ilyas H, van Hullebusch ED. Role of design and operational factors in the removal of pharmaceuticals by constructed wetlands. Water. 2019;**11**:2356. DOI: 10.3390/w11112356

Chapter 4

Pharmaceuticals in Wastewater: Current Issues and Future Trends for Their Treatment — A Critical Review

Alexis Rubén Bracamontes-Ruelas, Victor Samir Jaik-Morán and Lucia del Pilar Rodríguez-Hernández

Abstract

This chapter carries out a critical analysis of the literature review on the presence of pharmaceuticals in wastewater, their classification (e.g., Nonsteroidal Anti-inflammatory Drugs, antibiotics. etc.), their distribution in the environment, their toxicity, their endocrine-disrupting power, the antimicrobial resistance they may promote, and the most commonly used methods for detecting them in wastewater. Additionally, some of the most used physical, physicochemical, biological, and chemical alternatives (processes) for treating these emerging contaminants (pharmaceuticals) are evaluated and compared. Finally, in this chapter also the most effective process (treatment method) is determined and recommended as a treatment future trend based on criteria such as the process' capability to remove these compounds, operating costs process, the ease of scaling up of the treatment, the byproducts treatment generate (oxidation or degradation products), and treatment ability to adapt to conventional wastewater treatment processes (e.g., oxidation lagoons and activated sludge processes).

Keywords: wastewater, pharmaceuticals, toxicity, endocrine disruptors, antimicrobial resistance, antibiotics, physical treatment, physicochemical treatment, biological treatment, chemical treatment

1. Introduction

It has been demonstrated that the conventional processes currently used worldwide to treat wastewater are not able of removing many contaminants, among which emerging contaminants stand out [1, 2].

Emerging contaminants are natural or synthetic substances not subject to current legislation, and they are typically found in the environment at low concentrations ranging from nanograms to micrograms per liter [3]. These contaminants can be classified into groups, including pharmaceuticals, personal care products, hormones, pesticides, and more [2].

Pharmaceuticals are the most concerning group of emerging contaminants because they are biologically active compounds [4]. Furthermore, studies have evidenced that discharging these compounds into raw wastewater or effluents (treated wastewater) can cause mutagenic, toxicological and antimicrobial resistance issues for flora, fauna and human health, resulting in chronic ecological effects when there is a prolonged exposure [4, 5]. Likewise, the absence of more sensitive methods for detecting and monitoring these pharmaceuticals in wastewater and other aqueous matrices has exacerbated these issues.

Therefore, given these perspectives, this chapter provides a critical review of topics such as:

- The classification of pharmaceuticals.
- The distribution of pharmaceuticals in the environment.
- The toxicity that pharmaceuticals can cause.
- The endocrine-disrupting power of pharmaceuticals.
- The antimicrobial resistance that certain pharmaceutical groups (e.g., antibiotics) can cause.
- The most commonly used methods for detecting pharmaceuticals in wastewater.
- And finally, the presentation and selection of some used physical, physicochemical, biological or chemical treatment method to remove these compounds based on operating costs, ease of scaling, the by-products it generates (oxidation or degradation products), and its ability to adapt to conventional processes.

2. Classification of pharmaceuticals

The classification of pharmaceuticals is very diverse and depends primarily on the author. Nevertheless, the classification applied by Khasawneh and Palaniandy [6] in **Figure 1** is the one that best fits this research topic because it considers the following:

1. The potential environmental risk caused by each group of pharmaceuticals.
2. Their high consumption rate.
3. Their frequent detection in wastewater treatment plants (WWTPs), in both raw wastewater (influents) and effluents.

Figure 1 shows the main groups of pharmaceuticals considered by Khasawneh and Palaniandy [6]. These groups include antibiotics, lipid regulators, stimulants, Nonsteroidal Anti-inflammatory Drugs (NSAIDs), opioids, anti-diabetics, anticonvulsants, antidepressants, female hormones, and β-blockers.

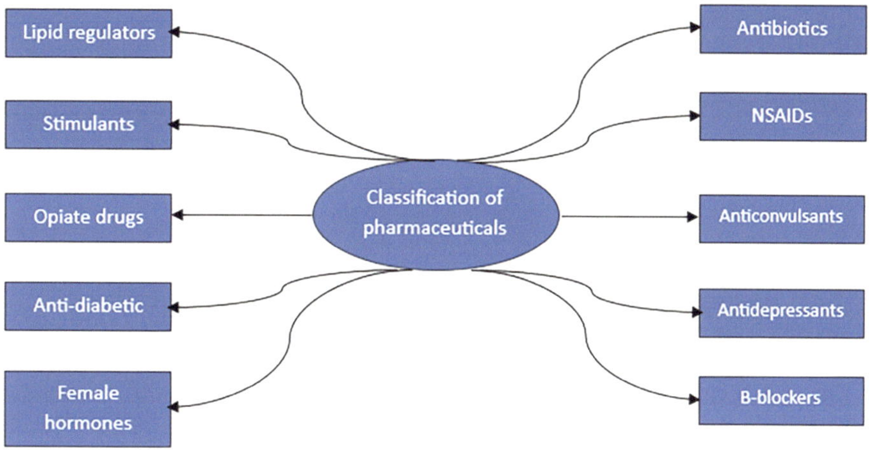

Figure 1.
Classification of pharmaceuticals [6].

It should be remarked, that according to published research, the focal group of pharmaceuticals that should be prioritized for treatment and removal in wastewater is antibiotics. These pharmaceuticals can generate antimicrobial (bacteria, fungi, parasites, and viruses) resistance (the topic is explored in greater depth in the following subsections) [7] in the environment and in wastewater treatment plants (WWTPs) because the concentrations that reach them are insufficient to eliminate microorganisms. Additionally, some pathogens excreted by people with diseases reach WWTPs along with antibiotics, which complicates the problem further since WWTPs allow for more direct interaction between antibiotic and microbes. This could cause major problems for human health in the future.

3. Distribution of pharmaceuticals in the environment

Multiple studies have shown that WWTPs are the key source of pharmaceuticals in the environment because conventional wastewater treatment processes (e.g., activated sludge and oxidation lagoons), which typically use microorganisms, are ineffective at removing this type of contaminants [8, 9]. Pharmaceuticals arrive at WWTPs in wastewater in both their original form (structure molecular) and as metabolites [10].

The distribution of each pharmaceutical in the environment generally depends on physicochemical properties such as:

- Its chemical and biological persistence.
- Its octanol-water partition coefficient ($\log K_{OW}$).
- Its volatility.
- The temperature and hydrogen potential (pH) level where the pharmaceutical is disposed in the environment.
- And the functional groups it contains (e.g., amines, amides, carboxyl, etc.) [11].

In this context, given the large number of pharmaceuticals and their diverse physicochemical properties, they can be found and distributed in almost all types of environmental compartments. This makes pharmaceuticals a problem of global importance [12].

4. Toxicity of pharmaceuticals

Regarding the toxicity of pharmaceuticals, various studies have demonstrated their harmful effects on ecosystems and human health, despite their low concentrations in the environment [13].

One of the groups of pharmaceuticals that is of particular concern to the scientific community is antibiotics, since, as mentioned above (Section. 2), this group has the ability to generate antimicrobial resistance in some pathogens [7, 14]. Moreover, it has also been found that some non-pathogenic microorganisms, which form part of the microbiota of environmental compartments, tend to be eliminated for this group of pharmaceuticals, altering the ecosystem, a factor that makes this group (antibiotics) a potential risk to the environment [14]. Nonetheless, the environmental damage caused by each antibiotic is directly related to its molecular structure and physicochemical properties [15].

Likewise, it has been found that groups of pharmaceuticals such as female hormones, present in WWTP effluents, can have endocrine-disrupting effects on both animals and humans. A clear example of this is the feminization and infertility in some fish species that has been caused when they are exposed to this kind of compounds [14–18].

Similarly, the following groups of pharmaceuticals can cause environmental risks due to their toxicity:

- Nonsteroidal Anti-Inflammatory Drugs (NSAIDs): according to the World Health Organization (WHO), NSAIDs can have harmful effects on both terrestrial and aquatic organisms, as well as on humans [19]. Studies have shown that NSAIDs have affected living organisms, causing chronic effects such as molecular, biochemical, or behavioral alterations [20], and although they are not yet considered a significant group of pharmaceuticals to be treated in wastewater, as scientists need to study the issue in greater depth, some research has shown that in the very near future they will be considered a relevant pharmaceutical group, given that some of these compounds tend to be absorbed by edible plants (e.g., watercress, barley, cucumber, tomato, lettuce, etc.) that are ingested by humans, leading to implicit and constant consumption of these compounds through the food chain [21].

- Anticonvulsants: prolonged exposure to these types of pharmaceuticals present in WWTP effluents has been shown to cause biochemical and behavioral changes in fish. These compounds accumulate in their tissues and are distributed and consumed by other living beings through the food chain [22].

- Antidepressants: studies have shown that antidepressants in WWTP effluents discharged into aquatic environments can alter the life cycles of aquatic organisms [23, 24]. These alterations can range from neurological to reproductive problems [24]. One clear example is the accumulation of these compounds in the livers and brains of fish and aquatic invertebrates [25].

- B-blockers: research has shown that β-blockers are consistently present in WWTPs. WWTP processes have been found to be ineffective at completely removing these compounds, making WWTP effluents the primary source of their entry into surface waters [26]. These compounds are toxic to aquatic organisms, such as fish, algae, and invertebrates, at low concentrations and can cause abnormal behavior in these species [26, 27]. Besides, concern in the scientific community has been raised because it is believed that they can reach drinking water supplies and potentially damage human health [27].

- Lipid regulators: scientific reports have found that prolonged exposure to these compounds causes damage to fish, including changes in swimming, feeding, and social behavior, although not all lipid regulators bioaccumulate [28]. Like other pharmaceutical groups, they primarily enter aquatic environments through raw sewage and WWTP effluents [29].

- Stimulants: this group of pharmaceuticals tends to be present in high concentrations in wastewater, including raw wastewater and effluents, which can cause serious environmental problems [30]. Stimulants can be toxic to aquatic organisms [31]. One of the most widely consumed stimulants is caffeine. Laboratory tests have revealed that this compound can have adverse effects on some coastal species by inhibiting their growth, affecting their reproduction and metabolic activity, and damaging their cells [32].

- Opiate drugs: these types of drugs have been found in raw wastewater and in effluents from WWTPs. Opioids are used by people worldwide, both with and without a prescription (i.e., illegally acquired). These compounds pose a latent risk to the environment because they can reach surface water, groundwater, and drinking water treatment plants [33]. Laboratory studies have shown that these drugs can be toxic to some organisms. For instance, codeine has been found to bioaccumulate in fish, reducing their hormone levels. Additionally, tramadol has been found to alter the behavior and mobility of crayfish [34]. This information reflects their toxicity. However, according to previous research, further studies are needed to better estimate their toxicity [33].

- Anti-diabetic: among the group of pharmaceuticals known as anti-diabetics, metformin is the most widely used compound [35]. Metformin has been consistently found in aquatic environments, has different toxicity levels in different organisms, and is possibly bio accumulative [36, 37]. It has even been shown that the presence of this compound in aquatic environments affects the entire food chain [35], which has caused concern among the scientific community.

Thus, based on the information compiled in this section, it is clear that more studies are needed to investigate the toxicity and harmful effects that certain groups of pharmaceuticals can have on human health and terrestrial biota. Plus, as previously mentioned, it is important to note that many of these pharmaceuticals are

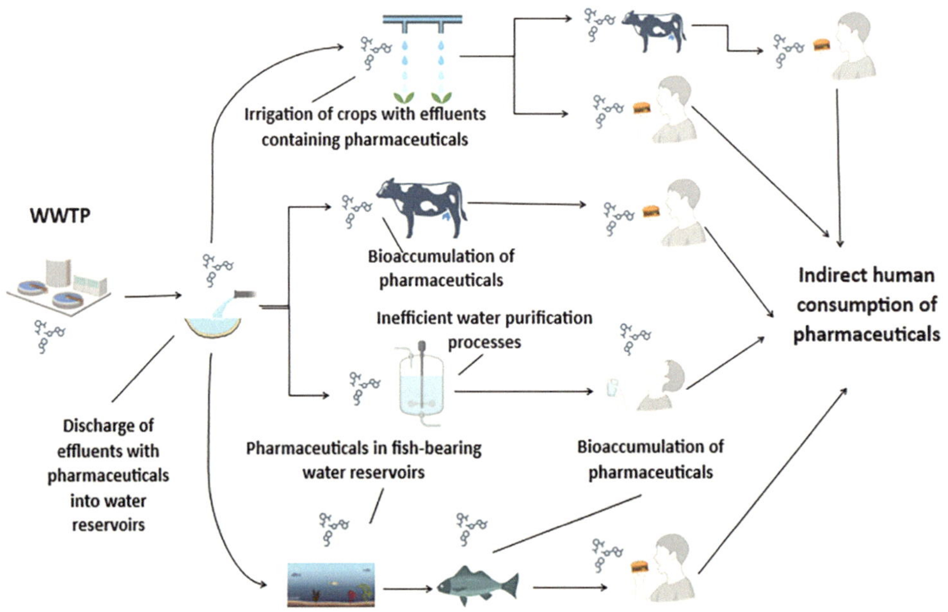

Figure 2.
General distribution of pharmaceuticals in the environment.

consumed by humans and animals through the food chain, with WWTPs being their main point of entry into food chain and environment (Section. 3).

Now, considering the factors influencing the distribution of pharmaceuticals in the environment, as discussed in Section. 3, and the data presented in this section, **Figure 2** offers a general overview of their distribution. It summarizes how pharmaceuticals can reach human consumption through environmental disposal and bioaccumulation with crops, terrestrial animals, and aquatic animals. Note that **Figure 2** only illustrates certain distribution routes that pharmaceuticals can take, and not all pharmaceuticals behave the same way because their physicochemical properties can vary significantly from compound to compound (Section. 3).

Finally, it should be emphasized that **Figure 2** should be modified for developing countries without the capacity to treat wastewater using WWTPs. In this case, the **Figure 2** should show that raw wastewater discharges are the main source of pharmaceuticals entering the environment rather than effluents from WWTPs.

5. Endocrine-disrupting power of pharmaceuticals

All compounds classified as endocrine disruptors are so designated because they can cause adverse effects in animals and humans by directly altering the endocrine system. These disruptors can block or mimic the natural hormones that are responsible for the proper functioning of certain bodily systems [38]. The most potent endocrine disruptors are female hormones such as estrone (E1), 17β-estradiol (E2), 17α-ethinylestradiol (EE2), and estriol (E3) [39, 40].

Various studies have found that, although endocrine disruptors such as female hormones are commonly detected at trace levels in bodies of water, these disruptors have the capacity to affect the endocrine functions of aquatic organisms. This can result in the feminization of male fish (Section. 4), reproductive problems, and alterations in sex ratios [40]. Also, studies have shown that 17β-estradiol can affect the transcription of the sex-related genes *Icam5*, *Lhb*, and *Fshb* during the neonatal period [41].

To end, it should be noted that the mechanisms by which endocrine disruptors act in animals and humans are not yet fully understood [38]. This indicates the need for further studies to improve the comprehension of these mechanisms. Nonetheless, additional studies should also be conducted to investigate the potential of other pharmaceutical classes to disrupt the endocrine system, given that research tends to focus on female hormones.

6. Pharmaceuticals and antimicrobial resistance problems

Among all pharmaceutical groups; antibiotics have been found to generate the highest level of antimicrobial resistance. This resistance has become a major problem in the 21st-century that can affect human health, animals, and environmental ecosystems [42].

Recently, antimicrobial resistance has worsened due to the indiscriminate release of antibiotics into the environment. This has created selective pressure that has driven the evolution of antibiotic-resistant microorganisms, such as bacteria, and facilitated the transfer of antibiotic resistance genes [43].

As mentioned in Section. 2, WWTPs are a focal point of utmost importance for the spread of antimicrobial resistance. Wastewater is often contaminated with antibiotics, microorganisms, antibiotic resistance genes, and other nutrients (organic and inorganic) that promote microbial growth, creating an ideal environment for the development of this problem. Additionally, the close proximity of microorganisms in WWTPs favors the dissemination and evolution of antibiotic resistance genes [42].

Figure 3 exhibits the possible sources of antibiotics, microbes, and resistance genes, as well as their path to WWTPs and subsequent distribution in the

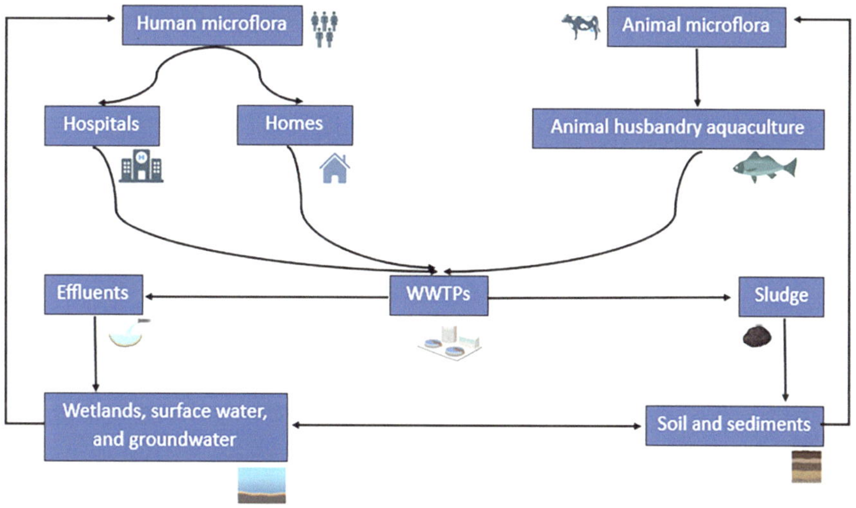

Figure 3.
Distribution of antibiotics, microorganisms, and resistance genes in the environment [44].

environment through wastewater. **Figure 3** shows that hospitals, homes, and aquaculture farms are the main sources of these compounds. These sources distribute antibiotics, microbes, and resistance genes through effluents and sludge generated by WWTPs. These effluents and sludge reach wetlands, groundwater, surface water, sediments, and ultimately, they return to human and animal consumption. It is worth mentioning that, in developing countries, raw wastewater rather than WWTPs is the main source of antibiotics, microbes, and resistance genes entering [44].

On the other hand, bacteria can transmit resistance genes through either vertical or horizontal evolution [44]:

- Vertical evolution: this type of evolution occurs through genetic variation among microbial populations that are subjected to pressures in their habitats which promote natural selection and the inheritance of such variation through progeny. For instance, a mutant gene inherited by certain bacteria due to antibiotic pressure, which alters the synthesis of proteins usually targeted by these compounds, is attributed to vertical evolution [44].

- Horizontal evolution: this evolution occurs through mechanisms that transfer resistance genes between identical, similar, or different bacterial species. **Figure 4** shows that resistance gene transfer can occur in three ways: conjugation, transduction, and transformation. These mechanisms can occur separately or simultaneously in bacteria to generate antibiotic resistance. In the case of

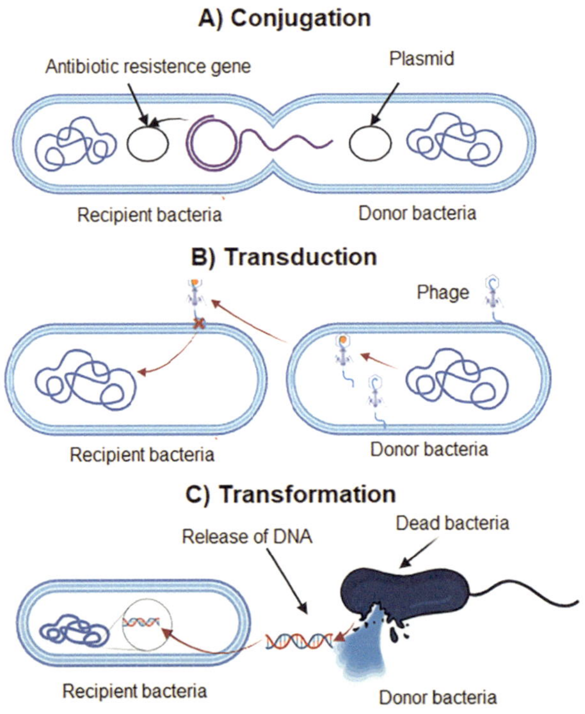

Figure 4.
Mechanisms of resistance gene transmission through horizontal evolution [44].

conjugation, direct contact is necessary to transmit resistance plasmids (resistance genes) from the donor bacteria to the recipient bacteria (**Figure 4**). In transduction, antibiotic resistance genes are transferred by bacteriophages (**Figure 4**). In transformation, a dead donor bacteria releases resistance plasmid, which are received by the recipient bacteria (**Figure 4**). Typically, transformation for gene transfer occurs in Gram-negative bacteria. For these gene transfer mechanisms to work, a bacterium that has developed resistance must act as the donor of the resistance gene or genes [44].

Consequently, having said the above, of the pharmaceutical groups discussed in this chapter, antibiotics in wastewater (treated and untreated) in conjunction with certain microorganisms (e.g., bacteria) and their gene transmission mechanisms (**Figure 4**) have caused antimicrobial resistance to evolve rapidly today, indicating that antimicrobial resistance is a latent problem and that physical, biological, or chemical processes must be proposed to remove them from wastewater, since according to various studies, antibiotic resistance in the years 1990 to 2021 has been directly linked to more than 1 million deaths per year worldwide [43].

7. Methods for detecting pharmaceuticals in wastewater

The most widely used method for detecting pharmaceuticals in wastewater today is liquid chromatography coupled to tandem mass spectrometry (LC–MS/MS), as it is a powerful method that enables us to detect pharmaceuticals at concentrations ranging from nanograms to micrograms per liter (ng/L-μg/L) [45, 46]. LC–MS/MS is also a popular choice for the determination of pharmaceuticals due to the high polarity exhibited by most compounds of this type [46].

Nevertheless, due to the complex nature of wastewater, this method (LC–MS/MS) requires prior concentration and purification techniques, such as solid-phase extraction (SPE). This is done to minimize possible interference and ensure proper pharmaceutical detection [46].

It should be noted that, although the LC–MS/MS method has been enhanced with purification and pretreatment techniques such as SPE, this method still presents problems such as negative identification of pharmaceuticals. In addition, it has been shown that during the SPE process, compounds present in wastewater with physicochemical properties similar to those of the pharmaceuticals to be extracted may also be retained in the SPE process, which can lead to low extraction recovery for such compounds (pharmaceuticals) [45].

Therefore, to improve the determination of pharmaceuticals, future research should focus on enhancing the efficiency and selectivity of extraction, as well as developing new materials and designs for determining trace concentrations of pharmaceuticals in complex matrices, such as wastewater [46]. Finally, despite current problems with SPE and LC–MS/MS, these methods are currently capable of detecting a wide variety of pharmaceuticals at different detection levels (LOD) in wastewater, as shown in **Table 1** [47].

Note that the detection limits (LOD) for pharmaceuticals in wastewater may vary from study to study (**Table 1**), as LC–MS/MS equipment may differ and may have different operating conditions.

Pharmaceutical	Limit of detection (LOD)	Pharmaceutical	Limit of detection (LOD)
Ibuprofen	0.59 μg/L	Naproxen	0.06 μg/L
Metformin	0.25 μg/L	Oxcarbazepine	0.06 μg/L
Sulfamethoxazole	0.02 μg/L	Ketoprofen	0.32 μg/L
Ampicillin	0.20 μg/L	Acetaminophen	0.05 μg/L
Ciprofloxacin	50 ng/ L	Trimethoprim	5.18 ng/L

Table 1.
Examples of LC–MS/MS detection limits (LOD) for different pharmaceuticals in wastewater in various studies [1, 47].

8. Treatment processes for removing pharmaceuticals in wastewater

The physical, physicochemical, biological, and chemical processes used for wastewater treatment have been extensively studied. However, as shown in **Table 2**, each has its advantages and disadvantages.

Chemical treatments are the most effective method for treating wastewater and removing dissolved contaminants, such as pharmaceuticals. Though, these processes are costly in terms of operation and reagents (**Table 2**).

Biological treatments, on the other hand, are usually easier to operate and have lower operating costs. Nevertheless, they are unlikely to remove pharmaceuticals such as antibiotics (**Table 2**). This is problematic when treating wastewater containing these types of compounds. Moreover, various studies have shown that biological processes do not mineralize emerging contaminants, such as pharmaceuticals, because co-metabolization occurs instead of metabolization. Their metabolic system develops based on other substrates found in greater quantities, such as ammonium, carbonate salts, carbohydrates, lipids, and proteins, which causes incomplete degradation of pharmaceuticals [48].

	Physical or physicochemical treatment	Biological treatment	Chemical treatment
Methods	Filtration, Adsorption, Air flotation, Extraction, Flocculation Sedimentation	Anaerobic Aerobic Activated sludge	Thermal oxidation (combustion) Chemical oxidation (e.g., advanced oxidation processes (AOPs)), Ion exchange Chemical precipitation
Advantages	Low cost of capital, Relatively safe, Easy to operate	Easy maintenance Relatively safe Elimination of the dissolved contaminants Easy to operate	High degree of treatment Elimination of the dissolved contaminants
Disadvantages	Volatile emissions, High energetic cost complex Maintenance	Volatile emissions Require elimination of residual sludge Susceptible to toxins or antibiotics	High costs of capital and operation Difficult operation

Table 2.
Advantages and disadvantages of different wastewater treatments [45].

Now then, with regard to physical and physicochemical wastewater treatments, it has been determined that contaminants only change phase [49] and, although they are removed from the wastewater, the problem persists in another medium. Besides, these processes are known to be highly selective, unlike chemical treatments.

It should be emphasized that among the treatments shown in **Table 2**, the most notable are advanced oxidation processes (AOPs), since these types of treatments generate hydroxyl radicals ($HO^{\bullet}$), which are highly reactive species with low selectivity [50]; in other words, they tend to attack compounds such as pharmaceuticals indiscriminately. For example, a recent study found that the advanced oxidation process Fenton could simultaneously remove pharmaceuticals such as ibuprofen and carbamazepine from wastewater at the laboratory level [51]. This demonstrated the efficiency of AOPs in removing contaminants such as pharmaceuticals from wastewater, although they cannot yet be scaled up to macro or industrial level [52], compared to some physical, physicochemical, and biological treatments.

Lastly, of the chemical treatments shown in this section (**Table 2**), AOPs appear to be the best option for treating pharmaceuticals in wastewater due to the drawbacks of other treatments (physical, physicochemical, and biological). Nonetheless, more research is needed to determine if AOPs can be coupled with conventional processes for macro or industrial wastewater treatment level and if the oxidation by-products are more toxic than the pharmaceuticals [2, 3, 52].

9. Conclusions

Based on the information presented in this chapter, the following conclusions can be drawn:

- The main way pharmaceuticals enter the environment is through wastewater, whether it is treated by WWTPs or not.

- The distribution of pharmaceuticals in the environment depends on their physicochemical properties (Section. 3). Given the wide variety of these compounds and their different properties, pharmaceuticals can be found in almost all environmental compartments.

- Some pharmaceuticals bioaccumulate in flora and fauna through various means, eventually reaching the human food chain (**Figure 2**).

- Antibiotics and female hormones should be given the highest priority for treatment and removal in wastewater because antibiotics are the main cause of antimicrobial resistance, and female hormones are highly endocrine-disrupting.

- The toxic effects of pharmaceutical groups such as NSAIDs, anticonvulsants, antidepressants, β-blockers, lipid regulators, stimulants, opioids, and antidiabetics should be studied in depth.

- The most promising method currently used for detecting pharmaceuticals in wastewater is liquid chromatography coupled to tandem mass spectrometry (LC–MS/MS). However, this method still has disadvantages relating to the previous solid-phase extraction (SPE) technique used for purification and concentration. Therefore, future research in this area should focus on improving the efficiency and selectivity of SPE, as well as developing new materials and designs, to enable the detection of pharmaceuticals in complex matrices, such as wastewater.

- Chemical treatments, specifically AOPs, show greater capacity to remove pharmaceuticals from wastewater than physical, physicochemical, and biological treatments due to their low selectivity. Nevertheless, these processes are not yet carried out on a large scale, and their adaptability to conventional wastewater treatments has not been proven. Additionally, it has not yet been determined whether the oxidation by-products they generate are toxic. This suggests that future research should address these issues to enable large-scale implementation of AOPs for the treatment of pharmaceuticals in wastewater, either as standalone treatments or as treatments adapted to conventional processes.

Acknowledgments

Acknowledgements to Dagoberto Rodríguez-Ortiz, Luis Arturo Torres-Castañon, José-Rafael Irigoyen-Campuzano, and Liliana Reynoso-Cuevas from CIMAV-DGO for their valuable technical support and collaboration.

Conflict of interest

The authors declare no conflict of interest.

Notes/thanks/other declarations

Thanks to my God, my entire family, and my friends who have always supported me in fulfilling my dreams. Especially Beatriz Ruelas (RIP), Hugo Ruelas (RIP), Ezequiel Soto, and Lucía Rodríguez.

Author details

Alexis Rubén Bracamontes-Ruelas[1*], Victor Samir Jaik-Morán[2] and Lucia del Pilar Rodríguez-Hernández[3]

1 Department of Sustainable Engineering, Advanced Materials Research Center, Durango, Mexico

2 Department of Industrial Engineering, Guadalupe Victoria, Higher Technological Institute of the Llanos Region, Durango, Mexico

3 The Equity, Chrisman, IL, United States

*Address all correspondence to: alexis.bracamontes@cimav.edu.mx

References

[1] Bracamontes-Ruelas AR, Ibarra-Rodríguez D, Rodríguez-Campos J, Velázquez-Fernández JB, Reyes-Vidal Y, Reynoso-Cuevas L. Evaluation of the presence of emerging contaminants in a municipal wastewater treatment plant in Durango, Mexico. Case Studies in Chemical and Environmental Engineering. 2025; **11**:1012218. DOI: 10.1016/j.cscee.2025.101218

[2] Bracamontes-Ruelas AR, Ordaz-Díaz LA, Bailón-Salas AM, Ríos-Saucedo JC, Reyes-Vidal Y, Reynoso-Cuevas L. Emerging pollutants in wastewater, advanced oxidationprocesses as an alternative treatment and perspectives. Processes. 2022;**10**. 1041. DOI: 10.3390/pr10051041

[3] Bracamontes-Ruelas AR, Irigoyen-Campuzano JR, Torres-Castañon LA, LA R-C. Application of Advanced Oxidation Processes for Domestic and Industrial Wastewater Treatment. In B KT, E J-L, M CD, Queiroz Zepka L, editors. Wastewater Treatment and Sludge Management Systems - the Gutter-to-Good Approaches. 1st ed. Croatia: IntechOpen; 2024. p. 1–19. DOI: 10.5772/intechopen.1004636

[4] Samal K, Mahaptra S, Hibzur Ali M. Pharmaceutical wastewater as Emerging Contaminants (EC): Treatment technologies, impact on environment and human health. Energy Nexus. 2022; **6**:100076. DOI: 10.1016/j.nexus.2022.100076

[5] Paíga P, Figueiredo S, Correia M, André M, Barbosa R, Jorge S, Delerue-Matos C. Occurrence of 97 pharmaceuticals in wastewater and receiving waters: analytical validation and treatment influence. Journal of Xenobiotics. 2025;**15**:78. DOI: 10.3390/jox15030078

[6] Khasawneh OFS, Palaniandy P. Occurrence and removal of pharmaceuticals in wastewater treatment plants. Process Safety and Environmental Protection. 2021;**150**:532–556. DOI: 10.1016/j.psep.2021.04.045

[7] Ahmed SK, Hussein S, Qurbani K, Hussein Ibrahim R, Fareeq A, Ali Mahmood K, Gamal Mohamed M. Antimicrobial resistance: Impacts, challenges, and future prospects. Journal of Medicine, Surgery, and Public Health. 2024;**2**:100081. DOI: 10.1016/j.glmedi.2024.100081

[8] Papageorgiou M, Kosma C, Lambropoulou D. Seasonal occurrence, removal, mass loading and environmental risk assessment of 55 pharmaceuticals and personal care products in a municipal wastewater treatment plant in Central Greece. Science of the Total Environment. 2016;**543**:547–569. DOI: 10.1016/j.scitotenv.2015.11.047

[9] Mishra RK, Mentha SS, Misra Y, Dwivedi N. Emerging pollutants of severe environmental concern in water and wastewater: A comprehensive review on current developments and future research. Water-Energy Nexus. 2023;**6**:74–95. DOI: 10.1016/j.wen.2023.08.002

[10] Rivera-Jaimes JA, Postigo C, Melgoza-Alemán RM, Aceña J, Barceló D, López de Alda M. Study of pharmaceuticals in surface and wastewater from Cuernavaca, Morelos,

Mexico: Occurrence and environmental risk assessment. Science of the Total Environment. 2018;**613-614**:1263–1274. DOI: 10.1016/j.scitotenv.2017.09.134

[11] Pereira Caixeta M. Contaminants of emerging concern: A review of risk assessment and treatment strategies. U.Porto Journal of Engineering. 2023;**9**:191–228. DOI: 10.24840/2183-6493_009-001_001282

[12] SanJuan-Reyes N, Gómez-Oliván LM, Islas-Flores H, Galar-Martínez M, García-Medina S, Dublán-García O, Hernández-Navarro MD. Occurrence of pharmaceuticalsin the environment. In: Gómez-Oliván L, editor. Ecopharmacovigilance. The Handbook of Environmental Chemistry. 1st ed. Chamb: Springer; 2017. p. 43–56. DOI: 10.1007/698_2017_142

[13] Nguyen MK, Lin C, Bui XT, Rakib MRJ, Nguyen HL, Truong QM, Hoang HG, Tran HT, Malafaia G, Idris AM. Occurrence and fate of pharmaceutical pollutants in wastewater: Insights on ecotoxicity, health risk, and state–of–the-art removal. Chemosphere. 2024; **354**:141678. DOI: 10.1016/j.chemosphere.2024.141678

[14] Estrada-Almeida AG, Castrejón-Godínez ML, Mussali-Galante P, Tovar-Sánchez E, Rodríguez A. Pharmaceutical pollutants: Ecotoxicological impacts and the use of agro-industrial waste for their removal from aquatic environments. Journal of Xenobiotics. 2024;**14**:1465–1518. DOI: 10.3390/jox14040082

[15] Fakhri BMS, Ghassemi Barghi N, Moradnia Mehdikhanmahaleh M, Raeis Zadeh SMM, Mousavi T, Rezaee R, Daghighi M, Abdollahi M. Pharmaceutical wastewater toxicity: An ignored threat to the public health. Sustainable Environment. 2024; **10**:2322821. DOI: 10.1080/27658511.2024

[16] Raju KV, Harini T, Kireeti VAR, Priva AS, Prasanthi NL. Impact of pharmaceutical toxicity on the environment and its regulatory aspects. Current Science. 2024;**126**:442–451. DOI: 10.18520/cs/v126/i4/442-451

[17] Kayode-Afolayan SD, Ahuekwe EF, Nwinyi OC. Impacts of pharmaceutical effluents on aquatic ecosystems. Scientific African. 2022;**17**: e01288. DOI: 10.1016/j.sciaf.2022.e01288

[18] Almazrouei B, Islayem D, Alskafi F, Catacutan MK, Amna R, Nasrat S, Sizirici B, Yildiz I. Steroid hormones in wastewater: Sources, treatments, environmental risks, and regulations. Emerging Contaminants. 2023; **9**:100210. DOI: 10.1016/j.emcon.2023.100210

[19] Mussa ZH, Al-Qaim FF, Jawad AH, Scholz M, Yaseen ZM A comprehensive review for removal of non-steroidal anti-inflammatory drugs attained from wastewaterobservations using carbon-based anodic oxidation process. Toxics. 2022;**10**:598. DOI: 10.3390/toxics10100598

[20] Stec M, Astel A. Occurrence and toxicological assessment of six non–steroidal anti–inflammatory drugs (NSAIDs) in a wastewater treatment plant in Słupsk (Poland). Ecohydrology & Hydrobiology. 2024;**24**:523–534. DOI: 10.1016/j.ecohyd.2024.04.007

[21] Rastogi A, Kumar Tiwiri MK, Ghangrekar MM. A review on environmental occurrence, toxicity and

microbial degradation of Non-Steroidal Anti-Inflammatory Drugs (NSAIDs). Journal of Environmental Management. 2021;**300**:113694. DOI: 10.1016/j.jenvman.2021.113694

[22] Ferencik M, Blahova J, Schovankova J, Siroka Z, Svobodova Z, Kodes V, Stepankova K, Lakdawala P. Residues of selected anticonvulsive drugs in surface waters of the Elbe River Basin (Czech Republic). Water. 2022;**14**:4122. DOI: 10.3390/w14244122

[23] Melchor-Martínez EM, Jiménez-Rodríguez MG, Martínez-Ruiz M, Peña-Benavides AP, Iqbal HMN, Parra-Saldívar R, Sosa-Hernández JE. Antidepressants surveillance in wastewater: Overview extraction and detection. Case Studies in Chemical and Environmental Engineering. 2021; **3**:100074. DOI: 10.1016/j.cscee.2020.100074

[24] Sun P, Liu H, Zhao Y, Hao N, Deng Z, Zhao W. A novel data-driven screening method of antidepressants stability in wastewater and the guidance of environmental regulations. Environment International. 2025; **198**:109427. DOI: 10.1016/j.envint.2025.109427

[25] Wang W, Zhang J, Hu M, Liu X, Sun T, Zhang H. Antidepressants in wastewater treatment plants: Occurrence, transformation and acute toxicity evaluation. Science of the Total Environment. 2023; **903**:16610. DOI: 10.1016/j.scitotenv.2023.166120

[26] Xu J, Sun H, Zhang Y, Alder AC. Occurrence and enantiomer profiles of β-blockers in wastewater and a receiving water body and adjacent soil in Tianjin, China. Science of the Total Environment. 2019;**650**:1122–1130. DOI: 10.1016/j.scitotenv.2018.09.086

[27] Yi M, Sheng Q, Sui Q, Lu H. b-blockers in the environment: Distribution, transformation, and ecotoxicity. Environmental Pollution. 2020; **266**:115269. DOI: 10.1016/j.envpol.2020.115269

[28] Blonç M, Lima J, Balasch JC, Tort L, Gravato C, Teles M. Elucidating the effects of the lipids regulators fibrates and statins on the health status of finfish species: A review. Animals. 2023;**13**:792. DOI: 10.3390/ani13050792

[29] Zhang K, Zhao Y, Fent K. Cardiovascular drugs and lipid regulating agents in surface waters at global scale: Occurrence, ecotoxicity and risk assessment. Science of the Total Environment. 2020; **729**:138770. DOI: 10.1016/j.scitotenv.2020.138770

[30] Egea-Corbacho A, Gutiérrez S, Coello D, Quiroga JM. Comparison in the removal of stimulants and antibiotics from wastewater for its subsequent reuse with different technologies. Chemical Engineering Research and Design. 2021;**166**:191–196. DOI: 10.1016/j.cherd.2020.12.006

[31] Styszko K, Dudarska A, Zuba D. The presence of stimulant drugs in wastewater from Krakow (Poland): A snapshot. Bulletin of Environmental Contamination and Toxicology. 2016;**97**:310–315. DOI: 10.1007/s00128-016-1869-5

[32] Vieira LR, Amvm S, Freitas R. Caffeine as a contaminant of concern: A review on concentrations and impacts in marine coastal systems. Chemosphere. 2022; **286**:131675. DOI: 10.1016/j.chemosphere.2021.131675

[33] Campos-Mañas MC, Ferrer I, Thurman EM, Agüera A. Opioid

occurrence in environmental water samples—A review. Trends in Environmental Analytical Chemistry. 2018;**20**:e00059. DOI: 10.1016/j.teac.2018.e00059

[34] Niroula V, Pagsuyoin SA. Stability and Degradation of Opioids in River Water. ACS Omega. 2024;**9**:26355–26362. DOI: 10.1021/acsomega.4c02486

[35] Balakrishnan A, Sillanpää M, Jacob MM, Vo DVN. Metformin as an emerging concern in wastewater: Occurrence, analysis and treatment methods. Environmental Research. 2022; **213**:113613. DOI: 10.1016/j.envres.2022.113613

[36] Zheng Y, Shao Y, Zhang Y, Liu Z, Zhao Z, Xu R, Ding J, Li W, Wang B, Zhang H. Metformin as an emerging pollutant in the aquatic environment: Occurrence, analysis, and toxicity. Toxics. 2024;**12**:483. DOI: 10.3390/toxics12070483

[37] Ambrosio-Albuquerque EP, Cusioli LF, Bergamasco R, Sinópolis Gigliolli AA, Lupepsa L, Paupitz BR, Barbieri PA, Borin-Carvalho LA, de Brito Portela-castro AL. Metformin environmental exposure: A systematic review. Environmental Toxicology and Pharmacology. 2021; **83**:103588. DOI: 10.1016/j.etap.2021.103588

[38] Kasonga TK, Coetzee MAA, Kamika I, Ngole-Jeme VM, Benteke Momba MN. Endocrine-disruptive chemicals as contaminants of emerging concern in wastewater and surface water: A review. Journal of Environmental Management. 2021; **277**:111485. DOI: 10.1016/j.jenvman.2020.111485

[39] Werkneh AA, Gebru SB, Redae GH, Tsige AG. Removal of endocrine disrupters from the contaminated environment: Public health concerns, treatment strategies and future perspectives - A review. Heliyon. 2022;**8**:e09206. DOI: 10.1016/j.heliyon.2022.e09206

[40] Latosińska J, Grdulska A. A review of methods for the removal of endocrine-disrupting compounds with a focus on oestrogens and pharmaceuticals found in wastewater. Applied Sciences. 2025;**15**:6514. DOI: 10.3390/app15126514

[41] Vilela CLS, Bassin JP, Peixoto RS. Water contamination by endocrine disruptors: Impacts, microbiological aspects and trends for environmental protection. Environmental Pollution. 2018;**235**:546–559. DOI: 10.1016/j.envpol.2017.12.098

[42] La Rosa MC, Maugeri A, Favara G, La Mastra C, Magnano San Lio R, Barchitta M, Agodi A. The impact of wastewater on antimicrobial resistance: A scoping review of transmission pathways and contributing factors. Antibiotics. 2025;**14**:131. DOI: 10.3390/antibiotics14020131

[43] Zhang S, Xu T, Zhang C, Zhou H, Liu T, Song H. A review of antibiotic resistance generation and control in wastewater and solid waste system. Journal of Environmental Chemical Engineering. 2025; **13**:118014. DOI: 10.1016/j.jece.2025.118014

[44] Mutuku C, Gazdag Z, Melegh S. Occurrence of antibiotics and bacterial resistance genes in wastewater: Resistance mechanisms and antimicrobial resistance control approaches. World Journal of Microbiology and Biotechnology. 2022;**38**:152. DOI: 10.1007/s11274-022-03334-0

[45] Fabregat-Safont D, Gracia-Marín E, Ibáñez M, Pitarch E, Hernández F. Analytical key issues and challenges in the LC-MS/MS determination of antibiotics in wastewater. Analytica Chimica Acta. 2023; **1239**:340739. DOI: 10.1016/j.aca.2022.340739

[46] Almeida CMM. Overview of sample preparation and chromatographic methods to analysis pharmaceutical active compounds in waters matrices. Separations. 2021;**8**:16. DOI: 10.3390/separations8020016

[47] Hernández-Tenorio R, Guzmán-Mar JL, Hinojosa-Reyes L, Ramos-Delgado N, Hernández-Ramírez A. Determination of pharmaceuticals discharged in wastewater from a public hospital using LC-MS/MS technique. Journal of the Mexican Chemical Society. 2021;**65**:94–108. DOI: 10.29356/jmcs.v65i1.1439

[48] Tran NH, Reinhard M, Gin KYH. Occurrence and fate of emerging contaminants in municipal wastewater treatment plants from different geographical regions-a review. Water Research. 2018;**133**:182–207. DOI: 10.1016/j.watres.2017.12.029

[49] Navidpour AH, Abbasi S, Li D, Mojiri A, Zhou JL. Investigation of advanced oxidation process in the presence of TiO_2 semiconductor as photocatalyst: Property, principle, kinetic analysis, and photocatalytic activity. Catalysts. 2023;**13**:232. DOI: 10.3390/catal13020232

[50] Cuerda-Correa EM, Alexandre-Franco MF, Fernández-González C. Advanced oxidation processes for the removal of antibiotics from water. An overview. Water. 2020;**12**:102. DOI: 10.3390/w12010102

[51] Bracamontes-Ruelas AR, Reyes-Vidal Y, Irigoyen-Campuzano JR, Reynoso-Cuevas L. Simultaneous oxidation of emerging pollutants in real wastewater by the advanced fenton oxidation process. Catalysts. 2023;**13**:748. DOI: 10.3390/catal13040748

[52] Kumar V, Lakkaboyana SK, Sharma N, Chakraborty P, Umesh M, Pasrija R, Thomas J, Kalebar VU, Jayaraj I, Awasthi MK, Das T, Oladipo AA, Barcelo D, Dumee LF. A critical assessment of technical advances in pharmaceutical removal from wastewater – A critical review. Case Studies in Chemical and Environmental Engineering. 2023; **8**:100363. DOI: 10.1016/j.cscee.2023.100363

Chapter 5

Tea, Teapots, and Toxicity in the Seventeenth and Eighteenth Century

Shirley Maloney Mueller and Michael Greenberger

Abstract

Lead levels in the exterior and interior of seventeenth and eighteenth-century Chinese export ceramics were tested. A steeped water technique was used to examine the interior. XRF standard testing was used for the exterior. The results indicated that lead was present in some of the teapots. This suggests that some tea ingested during this early period was laced with lead, thereby countering the contemporary belief that drinking tea in that time period was preferable to other alternatives.

Keywords: lead toxicity, early Chinese export teapots, seventeenth century, eighteenth century, steeped water technique, XRF

1. Introduction

Modern tea ingestion has been called 'immune system boosting' and 'healthy,' [1] among other positive attributes. Recently, however, attention has been drawn to the possible toxicity of modern tea consumption from exterior contamination. The kettle, heavy metals, and even the tea itself have been pointed to as culprits, among others.

In seventeenth- and eighteenth-century America, the attitude toward tea was quite similar to today's in that tea supposedly promoted health and was even therapeutic. However, we know that at least some of the tea imported from China to Europe was packed in lead-lined containers during its ocean voyage, which took up to two or more years [2, 3]. (**Figure 1**) In addition, the tea caddies in which the tea was stored at its destination were often lead-lined [4–9]. (**Figure 2**) While these were already dangerous, early pipes through which water flowed into household taps were lead-lined in some countries, such as England [10]. Lead contamination from these sources may have been minor, but there could be an accumulative impact.

2. Background for the study method

This study addresses this question using a unique analysis of lead within the pots by Dr. Shirley Maloney Mueller. This method will be referred to as the "steeped water method" for this discussion. The test reveals a way to look retrospectively for

IntechOpen

Figure 1.
Gouache, one of twelve, illustrating tea production. Chinese laborers are packing tea with their feet, often in a lead lined container. Chinese export product, 1803. Private collection, H. 17.5 in.; W. 21.3 in. Thomas M. Mueller photography, Indianapolis, Indiana, USA.

potential lead contamination within a teapot. While there are examinations in place to prove lead contaminates, such as X-ray fluorescence (XRF), these only show results of the exterior of objects. Dr. Mueller's new approach to testing lead can give us information about *inside* teapots as well.

A series of Chinese export teapots dated 1690 to 1790 were examined (Chart). Teapots such as these were used in the colonies [11]. In the 1685 inventory, colonist Dr. Jacob De Lange wrote:

> *"The variety of form among this useful and decorative collection indicates that every sort of porcelain enjoyed in Holland would have been available to the rich in New Netherland..."*

3. Methodology

Eight seventeenth and eighteenth-century Chinese export teapots (1690-1790) were tested for lead residual using the steeped water technique. This method is performed by boiling water in the pot, steeping it for 5 minutes, and then testing the water for lead contamination.

City tap water (Indianapolis, Indiana, USA) was boiled and then immediately poured into each pot. After 5 minutes, a water sample was transferred into special containers purchased from MACS Laboratory Inc. (Santa Clara, California).

Figure 2.
Pear shaped tea caddy similar to the lead lined caddies described in auction house records. England, eighteenth century. Private collection, H. 6.7 in.; W. 13.9 in. Thomas M. Mueller photography, Indianapolis, Indiana, USA.

The samples were then sent back to this laboratory for analysis. This test center is a commercial analyzer for lead that uses Environmental Protection Agency (United States) approved methods. It is accredited by the State of California Department of Health.

4. Results

The chart below summarizes the results. Six of the eight vessels exhibited significant levels of lead (>1.5 micrograms/liter). Two of these six showed *toxic lead levels* (>15 micrograms/liter). Another two of the eight demonstrated acceptable levels (< 1.5 micrograms/Liter) as did the control, which was examined using a contemporary glass teapot. More about the dating and characteristics of early teapots, including the eight in this study, can be found in earlier articles by Dr. Mueller [12–15].

The first steeped residual from the 1724 Davidson impaling Carr armorial porcelain teapot (**Figure 3**, left) tested at nearly twice the concentration of lead considered toxic (> 15 micrograms/liter), 26.9 micrograms/liter. Interestingly, the lead did wash out with a subsequent five-minute addition of boiling water. It was then reduced to today's acceptable environmental level (1.5 to 15 micrograms/liter), 3.8 micrograms/liter.

The 1690 Yixing teapot (**Figure 3**, right) first steeped residual tested for a lead level of 20.4, well above the 15 micrograms/liter deemed toxic. The level dropped to

Figure 3.
Chinese export porcelain dated 1724 on the left and a Yixing 1690 teapot on the right. Private collection. Left: H. 4.87 in; W. 7.0 in. Right: H. 4.2 in.; W. 5.0 in. Photos courtesy of the Indianapolis Museum of art, now Newfields, Indianapolis, Indiana USA.

Teapot	Date of teapot	Lead level 1st steeping micro grams/liter	Interpretation lead level
Davidson impaling Carr	1724	26.9	Toxic
Yixing Teapot	1690	20.4	Toxic
Blanc de Chine overdecorated in Holland	1690	2.8	High
Meissen inspired "Chinaman taking tea"	1740	5.2	High
Yixing, slip decorated	1740	7	High
Dated Anniversary for Norwegian market	1776	9.9	High
Double spouted vessel	1690	<1.5	Acceptable
American Flag	1790	<1.5	Acceptable
Contemporary Glass	2012	<1.5	Acceptable

Table 1.
Lead levels.

16.1 after a second addition of boiling water for 5 minutes, still above the poisonous level but lower than after the first steeping (**Table 1**). Yixing teapots were often left unwashed on the inside and the results of this "steeped water method" are in line with this concept.

Four teapots dated 1690 to 1776 demonstrated lead levels (1.5 to 15 micrograms/liter) which were less than toxic. They averaged 6.2 micrograms/liter and are listed below from the lowest to the highest.

1. Blanc de Chine over-decorated in Holland, circa 1690, 2.8 micrograms/liter.

2. Meissen inspired "Chinaman taking tea," circa 1740, 5.2 micrograms/liter.

3. Yixing, slip decorated, 1740, 7.0 micrograms/liter:

4. Dated Anniversary for the Norwegian market dated 1776, 9.9 micrograms/liter.

The final remaining two of the eight teapots measured below the environmental standard of lead, 1.5 micrograms/liter. They were the double-spouted vessel for tea or wine, 1690, and the American flag teapot, circa 1790.

To examine for lead on the exterior of the two pots that demonstrated toxic levels, Jennifer L. Mass, Ph.D., then Senior Conservation Scientist at Winterthur Museum in Delaware used X-ray fluorescence (XRF).

During XRF analysis, the surface of the object is bombarded with X-rays. The energy of these incoming X-rays is high enough to ionize atoms on the object's surface, causing them to give off their fluorescent X-rays. The energies of the fluorescent X-rays emitted by the object are characteristic of the atoms it contains, and so this technique can be used to determine the elemental composition of a material, such as an overglaze enamel (or in the case of the Yixing pot, the outside of the body).

The 1724 Davidson impaling Carr armorial porcelain teapot did demonstrate lead in its outside glaze, though much less than the overglaze enamels. The exterior of the 1690 Yixing teapot did not test positive for lead.

5. Discussion

Higher than acceptable lead levels by today's environmental standards were present in six of eight pots tested. Toxic levels of lead were present in two of these. The question is—what was the source of the lead?

Possibilities include residual tea contaminated with lead, evaporated water within the interior that was laced with lead, or, lastly, from the glaze within the porcelain teapots (the exterior of the Yixing teapot was lead-free). The latter possibility is unlikely for two reasons. One is that a high-fired glaze would not be expected to disperse into the solution at the relatively lower temperature of boiling water compared to the firing temperature used for the high-fired glaze. Secondly, the toxic level of lead in the 1724 Davidson impaling Carr teapot was lowered with a second steeping, indicating that the culprit responsible for the initial toxic level was washing out. This would not be expected if the lead level was due to the glaze within the teapot.

Further, in the 1724 armorial teapot, the base of the spout was strongly stained brown (**Figure 4**) and the glaze itself was eroded in this area. This wearing away could be because tea is acidic, and if left on a wet glaze repeatedly, it can be damaged. There was also dried brown matter initially in the bottom of the pot suggestive of remaining tea. Since the toxic lead level diminished with a second steeping, the residual tea was likely responsible for the initial excessive lead.

The Yixing teapot, on the other hand, is a different specimen altogether than the glazed porcelain pots. Though its exterior body did not contain lead, it is porous, unlike porcelain teapots which are impermeable when intact [15]. This porosity is consistent with the novelty that the Chinese ascribed to Yixing teapots. After a time, tea did not need to be added to the vessel to make tea because it permeates into the interior of the pot in earlier brews. This means that boiled water alone added to a Yixing pot (without tea) would result in tea released from the vessel's interior. The same could be true for any residual lead that was remaining from making lead-laced tea.

On the other end of the spectrum from toxic, two pots demonstrated lead levels within the acceptable environmental standard. They were the 1690 double-spouted pot and the 1790 American flag teapot. One explanation for the 1690 pot is that it was never used for tea, but instead for wine or condiments or not at all. There was no tea stain at the base of either of the two spouts.

Figure 4.
Arms of Auchmuty teapot. Chinese export porcelain, 1750. The three tea stained straining holes represented here are like those in the 1724 arms of Davis impaling Carr (the arms of Davis impaling Carr is unavailable for photography). Private collection, H. 5.12 in; W. 7.5 in. Thomas M. Mueller photography, Indianapolis, Indiana, USA.

The 1790 teapot, on the other hand, was used for tea. There were dark stains around and about the straining holes at the base of the spout that confirm this. Likely, this took place in the 1800s in America since the flag on the pot indicates it was destined for this country. Lead pipes were not installed in United States municipalities until the late nineteenth and early twentieth century. Thus, the water poured into the teapot may not have been lead-ridden unless it was used after the late 1800s. Also, the tea carried to America may not always have been stored in lead-lined chests or tea caddies this late.

6. Limitation of this study

There is a limitation of this study that could not be avoided. The teapots examined were executed in China 200 years or more ago. We do not know when they were used for tea during their timespan, if at all. What we can say is that pots that showed toxic lead levels did have lead contamination. Any adulteration likely would have been earlier in the teapot's history rather than later because as the vessels aged, they were perceived to be valuable and, therefore, likely not used.

7. Conclusion

Sources for lead poisoning in early America have been reported in multiple studies. They included pewter, pottery, crystal, paint, and solder [16]. The discovery of lead in exhumed bones from this period provides further documentation [17]. Now, tea itself, at least in some cases, can be added to the list, almost certainly due to contamination with lead from extraneous sources.

In summary, the results of this study suggest that early tea drinking did not necessarily counter the negative effects of drinking unboiled germ-laden fluids or alcohol. Instead, sometimes, it simply substituted one poison for another. How this made people ill during this period and how ill it made them can only be conjectured. However, Dr. Mueller's tests reveal that lead contamination was spread through drinking tea using a new methodology. Hopefully, these results will make way for more scientists and art historians to study the effects of lead in ceramics in the seventeenth and eighteenth centuries.

Author details

Shirley Maloney Mueller[1*] and Michael Greenberger[2]

1 Indiana University Department of Neurology, Indianapolis, Indiana, USA

2 Independent Art Historian, New York, USA

*Address all correspondence to: shirleymaloneymueller@gmail

References

[1] Naghma K, Hasan M. Tea and health: Studies in humans. Current Pharmaceutical Design. 2013;**19**(34):6141-6147. DOI: 10.2174/1381612811319340008

[2] Jorg CJA. Porcelain and the Dutch China Trade. The Hague: Martinus Nijhoff; 1982. p. 78

[3] Godden GA. Oriental Export Market Porcelain. London, Toronto, Sydney, New York: Granada; 1979. p. 28

[4] Sotheby's London, April 13, 2006, lot 339. A George I Britannia Standard Siler Tea Caddy Edward Gibbon, London. Sotheby's London; 1722, "removable lead lining"

[5] Sotheby's Melbourne, October 30-31, 2007, lot 209. A Burr Walnut Veneered Tea Caddy, circa 1730 with "original lead lined compartments"

[6] Sotheby's London, June 19, 1986, lot 43. A Pair of George III Silver Tea Caddies with Silver-Mounted Tortoise Shell Case (London, 1771, Mark T.I.), "one lead lined"

[7] Sotheby's London, December 4, 2013, lot 526. A George III Ivory Tea Caddy circa 1790 "with its original lead linings"

[8] Sotheby's New York, January 22, 20ll, lot 29. A George III Inlaid Mahogany Tea circa, England, circa 1790 with "lead lining"

[9] Sotheby's London, December 3, 2013, lot 366. A Group of Five Fruitwood, Fruit-Form Tea Caddies Late 18th/Early 19th Century, "all retaining remains of original lead lining underneath"

[10] Walski TM. A history of water distribution. American Water Works Association Journal. 2006;**98**(3):110-121

[11] Mudge JMC. Chinese Export Porcelain for the American Trade, 1785-1835. Newark: University of Delaware Press; 1962. pp. 98-100

[12] Mueller SM. Eighteenth-century Chinese export porcelain teapots: Fashion and uniformity. American Ceramic Circle Journal. 2005;**XIII**:5-16

[13] Mueller SM. 17th century teapots, imagination and diversity. Orientations. 2005;**37**(7):59-65

[14] Mueller SM. Revelations of the Ca Mau shipwreck: Chinese export porcelain teapots on the cusp. American Ceramic Circle Journal. 2009;**XV**:4-9

[15] Mueller SM. Lifting the lid: Early Chinese export teapots. Transactions of the Oriental Ceramic Society. 2006-2007;**71**:89-93

[16] Eubanks, Elsie Irene, "Lead poisoning from the colonial period to the present [thesis]", College of William & Mary - Arts & Sciences, USA, (1996). Paper 1539626037

[17] Aufderheide AC, Neiman FD, Wittmers LE Jr, Rapp G. Lead in bone II: Skeletal-lead content as an indicator of lifetime lead ingestion and the social correlates in an archaeological population. American Journal of Physical Anthropology. 1981;**55**:285-291